Lecture Notes in Mathematics

A collection of informal reports and seminars
Edited by A. Dold, Heidelberg and B. Eckmann, Zürich

Series: Mathematisches Institut der Un
Advisers: H. Bauer und K. Jacob

226

Seminar on Potential Theory, II

Edited by Heinz Bauer, Universität Erlangen-Nürnberg,
Erlangen/Deutschland

Springer-Verlag
Berlin · Heidelberg · New York 1971

AMS Subject Classifications (1970): 28 A 10, 30 A 78, 31 C 05, 31 C 15, 31 C 20, 31 C 25, 31 D 05

ISBN 3-540-05638-6 Springer-Verlag Berlin · Heidelberg · New York
ISBN 0-387-05638-6 Springer-Verlag New York · Heidelberg · Berlin

Offsetdruck: Julius Beltz, Hemsbach/Bergstr.

Contents

FUNCTIONAL SPACES AND THEIR EXCEPTIONAL SETS

by

Jürgen Bliedtner

0. Introduction

A. BEURLING and J. DENY developed in [1] the theory of Dirichlet spaces and treated potentials in such spaces. To get a more precise potential theory of a Dirichlet space $\mathcal{H} = \mathcal{H}(\Omega, m)$ they introduced a theory of capacity and replaced exceptional sets with respect to m-measure by sets of exterior capacity zero. The study of this associated capacity has been done by J. DENY [2] in the more general setting of a functional space.

In this paper we shall replace the inner product of such a functional space $\mathcal{H}$ by a continuous, coercive bilinear form a (see [5]) and study the potentials and capacity which are associated with a (§§ 2,3). Since the bilinear form a is not necessarily symmetric, we get a dual theory, i.e. we can introduce copotentials and a cocapacity. In § 3 we shall study the connection between the capacity and cocapacity. It will be shown in §§ 4,5 that this theory of capacity gives rise to Deny's class of quasi-continuous functions on Ω and his class of exceptional sets (the sets of exterior capacity zero).

In "DIRICHLET FORMS ON REGULAR FUNCTIONAL SPACES"(*) we shall construct examples for such bilinear forms from uniformly elliptic differential operators of order 2 which are not necessarily self-adjoint (see [6]), and shall show that nearly all potential-

(*) subsequent article

theoretic notions (as the domination principle, complete maximum principle, condensor principle etc.) in this axiomatic theory can be expressed by certain properties of the bilinear form a.

1. Coercive bilinear forms on Hilbert spaces

Let $\mathcal{H}$ be a real Hilbert space, the norm in $\mathcal{H}$ is denoted by $\|\cdot\|$, the associated scalar product by $\langle\cdot,\cdot\rangle$.
Let a be a real continuous bilinear form on $\mathcal{H} \times \mathcal{H}$ which is coercive, i.e. there exists a constant $c > 0$ such that

$$a(u,u) \geq c\|u\|^2$$

for all $u \in \mathcal{H}$.
G.STAMPACCHIA [5] proved the following representation theorem:

(1.1) THEOREM: If L is a real continuous linear form on $\mathcal{H}$ and A a non-empty, closed, convex subset of $\mathcal{H}$ then there exists a uniquely determined $u \in A$ such that

$$a(u,v-u) \geq L(v-u)$$

for every $v \in A$. If w represents L, i.e. $\langle w,x\rangle = L(x)$ for every $x \in \mathcal{H}$, the map $w \longmapsto u$ is continuous.

NOTATIONS: 1) If in theorem (1.1) L is of the form $L(v) = a(w,v)$ for a certain $w \in \mathcal{H}$, we call the unique point $u \in A$ the a-projection of w and write $u = \pi_A^a(w)$. For the adjoint bilinear form $\hat{a}$ which is defined by

$$\hat{a}(u,v) := a(v,u) \qquad (u,v \in \mathcal{H})$$

we write $\hat{\pi}_A^a$ instead of $\pi_A^{\hat{a}}$.

2) Application of theorem (1.1) to the whole space $\mathcal{H}$ yields that π^a and $\hat{\pi}^a$ are both algebraic and topological automorphisms of $\mathcal{H}$, where $a(\pi^a(w),v) = \langle w,v\rangle = a(w,\hat{\pi}^a(v))$.

3) In the special case that a is the scalar product

of $\mathcal{H}$, we shall omit the letter "a" in all definitions and notations.

The following approximation theorem of U. MOSCO can be found in [4]:

(1.2) THEOREM: Let $(A_i)_{i\in I}$ be an upward filtering family of non-empty closed convex subsets of $\mathcal{H}$. Then for any $w \in \mathcal{H}$, $\pi^a_{A_i}(w)$ converges strongly to $\pi^a_A(w)$, where $A = \overline{\bigcup_{i\in I} A_i}$.

2. Functional spaces and their pure potentials

In the following Ω will always denote a locally compact Hausdorff space and m a positive Radon-measure on Ω. By a function on Ω we always understand an extended real-valued m-measurable function on Ω. We do not distinguish notationally between a function and its class modulo locally m-negligible functions. We say that a property P holds a.e.[m] on an open subset ω of Ω if P holds on ω except on a set which is locally of zero m-measure.

For any open subset ω of Ω we denote by $M_c(\omega)$ the set of m-essentially bounded functions on Ω with compact support contained in ω. The support of a function f on Ω is defined as the complement of the largest open set $\omega \subset \Omega$ such that $f = 0$ a.e.[m] on ω. For any set $\mathcal{F}$ of functions, we define $\mathcal{F}^+ = \{f \in \mathcal{F}: f \geq 0 \text{ a.e.[m]}\}$.

Following J. DENY [2], we give the definition of a functional space:

DEFINITION: A functional space with respect to Ω and m is a Hilbert space $\mathcal{H} = \mathcal{H}(\Omega, m)$ whose elements are (classes of) locally m-integrable functions on Ω such that the following axiom holds:

(F) For any compact subset K of Ω, there exists a constant $A(K) > 0$ such that

$$\int_K |u|\,dm \leq A(K)\|u\|$$

for any $u \in \mathcal{H}$.

Let a be in the following a continuous coercive bilinear form on $\mathcal{H} \times \mathcal{H}$. An immediate consequence of the above definition is the following

(2.1) PROPOSITION: For each $f \in M_c(\Omega)$, there exists a unique element U_f^a (resp. $\hat{U}_f^a$) in $\mathcal{H}$ such that for any $v \in \mathcal{H}$

$$a(U_f^a, v) = \int vf\, dm \quad (\text{resp. } a(v, \hat{U}_f^a) = \int vf\, dm) .$$

The sets $\mathcal{H}_a = \{U_f^a : f \in M_c(\Omega)\}$ and $\hat{\mathcal{H}}_a = \{\hat{U}_f^a : f \in M_c(\Omega)\}$ are dense in $\mathcal{H}$.

Proof: By (F), the linear form

$$L(v) = \int vf\, dm$$

is continuous on $\mathcal{H}$, hence the first assertion follows from theorem (1.1) by taking $A = \mathcal{H}$.

Now assume that $w \in \mathcal{H}$ is orthogonal to $\mathcal{H}_a$. Then we have for every $f \in M_c(\Omega)$

$$0 = \langle U_f^a, w\rangle = a(U_f^a, \hat{\pi}^a(w)) = \int \hat{\pi}^a(w) f\, dm ,$$

whence $\hat{\pi}^a(w) = 0$ a.e.[m] on Ω and therefore $w = 0$ by the definition of $\hat{\pi}^a$. _/

DEFINITION: The element U_f^a (resp. $\hat{U}_f^a$) of (2.1) is called the a-potential (resp. a-copotential) generated by f. The number

$$a(U_f^a, U_f^a) = a(\hat{U}_f^a, \hat{U}_f^a)$$

is the a-energy of f.

The elements of the closed convex cone

$$\mathcal{P}^a = \overline{\{U_f^a : f \in M_c^+(\Omega)\}}$$

are called pure a-potentials. More generally, we define for any open set ω of Ω

$$\mathcal{P}_\omega^a = \overline{\{U_f^a : f \in M_c^+(\omega)\}} .$$

Evidently, $\mathcal{P}^a_\omega$ is a closed convex subcone of $\mathcal{P}^a_\Omega = \mathcal{P}^a$. The closed convex cones $\hat{\mathcal{P}}^a_\omega$ are defined in an analogous manner.

The next theorem gives an important characterization of pure a-potentials and more generally of the elements of $\mathcal{P}^a_\omega$:

(2.2) THEOREM: Let ω be an open subset of Ω. Then a necessary and sufficient condition for an element $u \in \mathcal{H}$ to be in $\mathcal{P}^a_\omega$, is

$$a(u,v) \geq 0 \quad \text{for all} \quad v \in \mathcal{H} \quad \text{such that} \quad v \geq 0 \quad \text{on } \omega .$$

Proof: Let $u \in \mathcal{H}$ and let u' be the $\mathring{a}$-projection of u on $\mathcal{P}^a_\omega$. Since $\mathcal{P}^a_\omega$ is a closed convex cone, u' is characterized by the following two properties:

$$(1) \quad a(u',u') = a(u',u) \quad ,$$

$$(2) \quad a(w,u') \geq a(w,u) \qquad \text{for all} \quad w \in \mathcal{P}^a_\omega .$$

Choosing $w = U^a_f$ with $f \in M^+_c(\omega)$, the inequality (2) yields

$$\int u' \, f \, dm = a(U^a_f,u') \geq a(U^a_f,u) = \int u \, f \, dm ,$$

hence $u' \geq u$ a.e.[m] on ω .

Now suppose $u \in \mathcal{H}$ satisfies the condition of the theorem. Since $u' - u \geq 0$ on ω, we have $a(u,u' - u) \geq 0$ and therefore by (1):

$$0 \leq c.\|u'-u\|^2 \leq a(u'-u,u'-u) = a(u',u'-u) - a(u,u'-u) \leq 0 \quad ,$$

whence $u = u' \in \mathcal{P}^a_\omega$.

The converse is evident for $u = U^a_f \in \mathcal{P}^a_\omega$ since

$$a(U^a_f,v) = \int vf \, dm \geq 0.$$

By the continuity of the bilinear form a, we get the desired inequality for $u \in \mathcal{P}^a_\omega$.

3. Capacity and cocapacity

For every open subset ω of Ω we define

$$\mathcal{K}_\omega = \{v \in \mathcal{K} : v \geq 1 \ \text{a.e.}[m] \ \text{on} \ \omega\}.$$

Obviously, $\mathcal{K}_\omega$ is a closed convex subset of $\mathcal{K}$. If $\mathcal{K}_\omega$ is non-empty, let $u_\omega = \pi^a_{\mathcal{K}_\omega}(0)$ and $\hat{u}_\omega = \hat{\pi}^a_{\mathcal{K}_\omega}(0)$.

DEFINITION: We define the a-capacity of an open set $\omega \subset \Omega$ as the number

$$\text{cap}_a\omega := \begin{cases} +\infty & , \text{ if } \mathcal{K}_\omega = \emptyset \\ a(u_\omega,u_\omega), & \text{ if } \mathcal{K}_\omega \neq \emptyset \end{cases} .$$

The a-cocapacity $\widehat{\text{cap}}_a\omega$ is defined by

$$\widehat{\text{cap}}_a\omega := \text{cap}_{\hat{a}}\omega .$$

(3.1) REMARKS: 1) u_ω is a pure a-potential, called the a-capacitary potential. Indeed, u_ω is characterized by (1.1) to be the only element in $\mathcal{K}_\omega$ such that for all $w \in \mathcal{K}_\omega$

$$a(u_\omega, w-u_\omega) \geq 0 .$$

If $v \in \mathcal{K}^+$, then $u_\omega + v \in \mathcal{K}_\omega$, hence

$$a(u_\omega,v) = a(u_\omega,(u_\omega+v) - u_\omega) \geq 0$$

and therefore $u_\omega \in \mathcal{P}^a$ by (2.2).

In an analogous manner, one shows that the a-cocapacitary potential $\hat{u}_\omega$ is a pure a-copotential.

2) $\text{cap}_a\omega = 0$ if and only if $m(\omega) = 0$. Indeed, $\text{cap}_a\omega = 0$ implies $u_\omega = 0$. Since $u_\omega \geq 1$ a.e.[m] on ω, we obtain $m(\omega) = 0$.

If $m(\omega) = 0$, then $0 \in \mathcal{K}_\omega$ and therefore $\text{cap}_a\omega = a(0,0) = 0$.

The following characterization of the a-capacity of an open set

ω in Ω is known for the classical Newtonian capacity as Kelvin's principle:

For an open set $\omega \subset \Omega$, the set

$$\mathcal{P}_\omega^{a,1} = \overline{\{U_f^a : f \in M_c^+(\omega) \text{ and } \int f \, dm = 1\}}$$

is a closed, convex subset of $\mathcal{H}$. If $\mathcal{P}_\omega^{a,1} \neq \emptyset$, then there exists by (1.1) a unique element $u'_\omega \in \mathcal{P}_\omega^{a,1}$ such that

$$a(v-u'_\omega, u'_\omega) \geq 0 \qquad \text{for all } v \in \mathcal{P}_\omega^{a,1},$$

i.e. u'_ω is the $\hat{a}$-projection of 0 onto $\mathcal{P}_\omega^{a,1}$.

Define:

$$\mathrm{CAP}_a\omega := \begin{cases} 0 & , \text{ if } \mathcal{P}_\omega^{a,1} = \emptyset \\ \dfrac{1}{a(u'_\omega, u'_\omega)} & , \text{ if } \mathcal{P}_\omega^{a,1} \neq \emptyset \end{cases} .$$

(3.2) THEOREM: For all open subsets $\omega \subset \Omega$,

$$\mathrm{cap}_a\omega = \mathrm{CAP}_a\omega .$$

Moreover, if $\mathrm{cap}_a\omega < +\infty$, then

$$u_\omega = \begin{cases} 0 & , \text{ if } m(\omega) = 0 \\ \dfrac{1}{a(u'_\omega, u'_\omega)} \cdot u'_\omega, & \text{ if } m(\omega) > 0 \end{cases} .$$

Proof: If $m(\omega) = 0$, then by remark 2 of (3.1), $\mathrm{cap}_a\omega = 0$. On the other hand, $\mathcal{P}_\omega^{a,1} = \emptyset$, hence $\mathrm{CAP}_a\omega = 0$. Assume now $m(\omega) > 0$. Since $u' := u'_\omega$ exists, we have

$$a(U_f^a - u', u') \geq 0$$

for all $f \in M_c^+(\omega)$ such that $\int f \, dm = 1$, hence

$$\int a(u',u') f \, dm = a(u',u') \leq a(U_f^a, u') = \int u' f \, dm$$

and therefore

$$(*) \qquad a(u',u') \leq u' \quad \text{a.e.}[m] \text{ on } \omega .$$

If $\mathrm{cap}_a\omega = +\infty$, there exists no element in $\mathcal{H}$ which is a.e.[m] ≥ 1 on ω. By (*) therefore $a(u',u') = 0$, whence $\mathrm{CAP}_a\omega = +\infty$. If $\mathrm{cap}_a\omega < +\infty$, then the a-capacitary potential $u := u_\omega$ of ω exists. Since $u \geq 1$ a.e.[m] on ω, we have

$$a(U_f^a, u) = \int uf\,dm \geq 1$$

for all $f \in M_c^+(\omega)$ such that $\int f\,dm = 1$.

Since u' is the limit of such U_f^a, we get $a(u',u) \geq 1$ which implies especially $u' \neq 0$ and therefore $\mathrm{CAP}_a\omega < +\infty$. Set $w := \dfrac{u'}{a(u',u')}$. By (*), $w \geq 1$ a.e.[m] on ω, which implies

$$a(u, w-u) \geq 0$$

by the definition of u.

Since $a(u',u) \geq 1$, we get $a(w,u-w) \geq 0$:

$$a(w,u-w) = a(w,u) - a(w,w) = \frac{a(u',u)}{a(u',u')} - \frac{1}{a(u',u')} \geq 0.$$

Hence $0 \leq c\|u-w\|^2 \leq a(u-w,u-w) = a(u,u-w) - a(w,u-w) \leq 0$, thus $u = w$ and therefore

$$\mathrm{cap}_a\omega = a(u,u) = a(w,w) = \frac{1}{a(u',u')} = \mathrm{CAP}_a\omega\ .$$ ⌟

The following example shows that in general the a-capacity and a-cocapacity of an open set do not coincide:

(3.3) <u>EXAMPLE</u>: If $\Omega = \{1,2\}$, $m = \varepsilon_1 + \varepsilon_2$ (unit mass in 1 and 2) then $\mathcal{H} = \mathbb{R}^2$ with the usual inner product is a functional space.

Define for $u = (u_1,u_2)$, $v = (v_1,v_2) \in \mathcal{H}$

$$a(u,v) = 2u_1v_1 + 4u_2v_2 - 3u_1v_2\ .$$

a is continuous and coercive with a constant $c = \frac{1}{2}$.

A simple calculation shows that the a-capacitary potential of Ω is given by $u = (1,1)$ and the a-cocapacitary potential of Ω by $\hat{u} = (\frac{3}{2},1)$. Therefore

$$\text{cap}_a\Omega = 3, \quad \text{but} \quad \widehat{\text{cap}}_a\Omega = 4 .$$

DEFINITION [3]: The projection T of the real line $\mathbb{R}$ on the closed interval [0,1] is called the unit contraction. We say that the unit contraction operates on $\mathcal{H}$ with respect to a if, for any $u \in \mathcal{H}$, the function $T \circ u$ is in $\mathcal{H}$ and the inequality

$$a(u+T\circ u, u-T\circ u) \geq 0$$

holds.

(3.4) LEMMA: Let $\omega \subset \Omega$ be open such that $\text{cap}_a\omega < +\infty$. If the unit contraction operates on $\mathcal{H}$ with respect to a, then the a-capacitary potential u of ω satisfies the following conditions:

(1) $0 \leq u \leq 1$ a.e.[m] on Ω.

(2) $u = 1$ a.e.[m] on ω.

Proof: By the definition of u, we have

$$a(u,v-u) \geq 0 \text{ for all } v \in \mathcal{H}_\omega = \{v \in \mathcal{H} : v \geq 1 \text{ a.e.}[m] \text{ on } \omega\}.$$

To show: $u = T\circ u$.

Let $u' := T\circ u$. Then $u' \in \mathcal{H}_\omega$ and $a(u,u'-u) \geq 0$.

Since T operates on $\mathcal{H}$ with respect to a, we have

$$a(u+u',u-u') \geq 0 \quad \text{or} \quad a(u',u-u') \geq a(u,u'-u),$$

therefore

$$0 \leq c\cdot\|u-u'\|^2 \leq a(u,u-u') - a(u',u-u') \leq 2a(u,u-u') \leq 0,$$

hence $u = u'$. $\_/$

(3.5) THEOREM: If the unit contraction operates on $\mathcal{H}$ with respect to a and $\hat{a}$, then for all open sets ω of Ω

$$\text{cap}_a\omega = \widehat{\text{cap}}_a\omega.$$

Proof: We may assume that $\mathcal{H}_\omega \neq \emptyset$ so that the a-capacitary and a-cocapacitary potential u and $\hat{u}$ exist. By remark 2 of (3.1) we may assume furthermore $m(\omega) > 0$. Let $u' \in \mathcal{P}_\omega^{a,1}$ such that by (3.2) $u = \frac{1}{a(u',u')} u'$.

It suffices to prove $a(u',\hat{u}) \leq 1$ since

$$\widehat{cap}_a \omega = a(\hat{u},\hat{u}) \leq a(u,\hat{u}) = \frac{1}{a(u',u')} \cdot a(u',\hat{u}) \leq cap_a \omega.$$

By the definition of u', for every $\varepsilon > 0$, there exists $f \in M_c^+(\omega)$ with $\int f\, dm = 1$ such that

$$a(u',\hat{u}) < a(U_f^a,\hat{u}) + \varepsilon.$$

Since $0 \leq \hat{u} \leq 1$, we get

$$a(U_f^a,\hat{u}) = \int f \hat{u}\, dm \leq \int f\, dm = 1,$$

hence $a(u',\hat{u}) \leq 1$. _/

Let α (resp. β) be the <u>symmetric</u> (resp. <u>anti-symmetric</u>) <u>part</u> of a, i.e.

$$\alpha(u,v) := \frac{a(u,v)+a(v,u)}{2}, \qquad \beta(u,v) := \frac{a(u,v)-a(v,u)}{2}$$

and let $A := \inf \{\alpha(u,u) : u \in \mathcal{H} : \|u\| = 1\}$, $B := \sup \{\beta(u,v) : u, v \in \mathcal{H} : \|u\| = \|v\| = 1\}$. For a proof of the following theorem, see G. STAMPACCHIA [6], p. 21o:

(3.6) <u>THEOREM</u>: Let $\omega_1, \omega_2 \subset \Omega$ be open such that $\omega_1 \subset \omega_2$. Then

$$cap_a \omega_1 \leq (1+\tfrac{B}{A})^2\, cap_a \omega_2 .$$

(3.7) <u>REMARKS</u>: (1) the a-capacity is monotone on the open sets of Ω if a is symmetric.

(2) In example (3.3) the a-capacitary potential of $\{2\}$ is $u = (0,1)$ and therefore $cap_a\{2\} = 4$, but $cap_a \Omega = 3$.

(3.8) <u>THEOREM</u>: If the unit contraction operates on $\mathcal{H}$ with respect to a and $\hat{a}$ then the relation $\omega_1 \subset \omega_2$ for open sets $\omega_1, \omega_2 \subset \Omega$ implies

$$cap_a \omega_1 \leq cap_a \omega_2 .$$

Proof: We may assume $\mathcal{K}_{\omega_2} \neq \emptyset$. Let u_1 (resp. u_2) be the a-capacitary (resp. a-cocapacitary) potential of ω_1 (resp. ω_2). Since $u_2 \geq 1$ a.e.[m] on ω_1, we have

$$a(u_1, u_2 - u_1) \geq 0.$$

Furthermore, we may assume $m(\omega_2) > 0$. Let $u' \in \hat{\mathcal{P}}^{a,1}_{\omega_2}$ such that by (3.2) $u_2 = \frac{1}{a(u',u')} u'$.

Since $0 \leq u_1 \leq 1$, we get as in the proof of (3.2)

$$a(u_1, u') \leq 1,$$

hence by (3.2) and (3.5)

$$\mathrm{cap}_a \omega_1 = a(u_1, u_1) \leq a(u_1, u_2) = \frac{1}{a(u',u')} \cdot a(u_1, u')$$

$$\leq \widehat{\mathrm{cap}}_a \omega_2 = \mathrm{cap}_a \omega_2. \quad \_/$$

4. Quasi-continuity

DEFINITION: An extended real-valued function f on Ω is called a-quasi-continuous if for any $\varepsilon > 0$, there exists an open set $\omega \subset \Omega$ such that

i) $\mathrm{cap}_a \omega < \varepsilon$;

ii) the restriction of f to $\complement\omega$ is continuous on $\complement\omega$.

The following estimations yield the independence of the notions of quasi-continuity from the special bilinear form a.

(4.1) THEOREM: For any open subset $\omega \subset \Omega$ the following estimates hold:

$$c \cdot \mathrm{cap}\ \omega \leq \mathrm{cap}_a \omega \leq \frac{C^2}{c} \mathrm{cap}\ \omega,$$

where C is the norm of the bilinear form a.

Proof: We may assume that $\mathcal{K}_\omega \neq \emptyset$ and $m(\omega) > 0$. Let u (resp. u_a) be the capacitary (resp. a-capacitary) potential of ω.

By the definition of u and u_a we get

$$(1)\quad \|u\| \leq \|u_a\|$$

$$(2)\quad a(u_a, u-u_a) \geq 0.$$

The first inequality follows by (1):

$$c \cdot \mathrm{cap}\ \omega = c \cdot \|u\|^2 \leq c \cdot \|u_a\|^2 \leq a(u_a, u_a) = \mathrm{cap}_a \omega\ ,$$

the second one by (2) and the coerciveness of a:

$$a(u_a, u_a) \leq a(u_a, u) \leq C\|u_a\| \cdot \|u\| \leq C \cdot c^{-\frac{1}{2}}\ a(u_a, u_a) \cdot \|u\|,$$

hence

$$\mathrm{cap}_a \omega = a(u_a, u_a) \leq \frac{C^2}{c} \|u\|^2 = \frac{C^2}{c} \mathrm{cap}\ \omega\ . \quad \_/$$

(4.2) <u>COROLLARY</u>: An extended real-valued function f on Ω is a-quasi-continuous iff f is quasi-continuous.

5. Exterior capacity and exceptional sets

For every subset $E \subset \Omega$, we define a closed, convex subset $\mathcal{H}_E \subset \mathcal{H}$ by

$$\mathcal{H}_E = \overline{\bigcup_{\substack{E \subset \omega \\ \omega \text{ open}}} \mathcal{H}_\omega}\ .$$

If $\mathcal{H}_E \neq \emptyset$, there exists a unique element $u_E \in \mathcal{H}_E$ by (1.1) such that

$$a(u_E, v-u_E) \geq 0 \quad \text{for all} \quad v \in \mathcal{H}_E.$$

u_E is a pure a-potential, called the <u>exterior a-capacitary potential</u> of E.

<u>DEFINITION</u>: The exterior a-capacity of a subset $E \subset \Omega$ is defined as

$$\mathrm{cap}_a^* E := \begin{cases} +\infty & , \text{ if } \mathcal{H}_E = \emptyset\ ; \\ a(u_E, u_E) & , \text{ if } \mathcal{H}_E \neq \emptyset\ . \end{cases}$$

(5.1) <u>REMARK</u>: If $\mathcal{H}_E \neq \emptyset$ then the family $(\mathcal{H}_\omega)_{E\subset\omega,\ \omega \text{ open}}$ is upward filtering, and by (1.2), u_E is the strong limit of a-capacitary potentials u_ω.

This remark and (4.1) yield at once the following

(5.2) <u>COROLLARY</u>: For any subset $E \subset \Omega$, the following estimates hold:

$$c \cdot \text{cap}^* E \leq \text{cap}_a^* E \leq \frac{C^2}{c} \cdot \text{cap}^* E \ .$$

<u>DEFINITION</u>: A set $E \subset \Omega$ is called an <u>a-quasi-null</u> set if

$$\text{cap}_a^* E = 0 .$$

(5.3) <u>COROLLARY</u>: The class of <u>a-quasi-null</u> sets is independent of the bilinear form a.

6. Bibliography

A.BEURLING, J.DENY :

[1] Dirichlet spaces. Proc.Nat.Acad.USA 45 (1959), 208-215.

J.DENY :

[2] Théorie de la capacité dans les espaces fonctionnels. Séminaire BRELOT-CHOQUET-DENY (Théorie du Potentiel) 9e année, 1964/65, no. 1.

M.ITO :

[3] A note on extended regular functional spaces. Proc. Jap. Acad. 43 (1967), 435-440.

U.MOSCO :

[4] Approximation of the solution of some variational inequalities. Ann.Sc.Norm.Sup.Pisa 21 (1967), 373-394.

G.STAMPACCHIA:

[5] Formes bilinéaires coercitives sur les ensembles convexes. C.R.Acad.Sc.Paris 258 (1964), 4413-4416.

[6] Le problème de Dirichlet pour les équations elliptiques du second ordre à coefficients discontinus. Ann.Inst.Fourier 15 (1965), 189-259.

Contents

DIRICHLET FORMS ON REGULAR FUNCTIONAL SPACES

by

Jürgen Bliedtner*)

0. Introduction

A. BEURLING and J. DENY developed in [2] the theory of Dirichlet spaces and treated potentials in such spaces. Since the associated kernel is always symmetric, M. ITO extended this theory by replacing the inner product of a regular functional space $\mathcal{H} = \mathcal{H}(\Omega, m)$ (in the sense of [5]) by a continuous bilinear form a on $\mathcal{H} \times \mathcal{H}$ which satisfies

$$(*) \qquad a(u,u) = \|u\|^2 \qquad (u \in \mathcal{H}).$$

He announced in [9] several conditions on a which are equivalent to the domination principle and the complete maximum principle. In the case of the complete maximum principle, his condition on a reads as follows:

$$(**) \qquad a(u+T_I \circ u,\ u-T_I \circ u) \geq 0 \quad \text{and} \quad a(u-T_I \circ u,\ u+T_I \circ u) \geq 0 \qquad (u \in \mathcal{H})$$

where T_I is the projection of $\mathbb{R}$ onto the closed unit interval.

In this paper, we shall replace the condition (*) by the following one: There exists a constant $c > 0$ such that $a(u,u) \geq c\|u\|^2 (u \in \mathcal{H})$ i.e. a is coercive in the sense of G. STAMPACCHIA [11]. The capacity theory associated to such a bilinear form a has been studied in [3].

This paper is organized as follows: After some preliminaries on coercive bilinear forms (§ 1) and the general potential

*) Partially supported by the National Science Foundation (Grant GP-13070).

theory connected with them (§§ 2-5), we study the domination principle (§ 6) and the complete maximum principle (§ 8). The latter is satisfied iff a is a Dirichlet form, i.e. a satisfies (**). In § 7 we introduce the kernel and the singular measure associated to a. As we shall see (§ 9), this singular measure plays an important role in the representation of a Dirichlet form. In § 10 we give the main example of a Dirichlet form using results of G. STAMPACCHIA [12]

In the second part (§§ 11-14) we shall prove that most of the potential-theoretic principles hold in the present situation. The principles are known in classical potential theory and are obtained in the case of Dirichlet spaces by A.BEURLING, J.DENY, M.ITO [2], [4], [6],[7],[8]. Finally, a theorem concerning the support of the balayaged measures is applied to the example of § 10.

I. Characterizations of Dirichlet forms

1. Coercive bilinear forms on Hilbert spaces

Let $\mathcal{H}$ be a real Hilbert space, the norm in $\mathcal{H}$ is denoted by $\|\cdot\|$, the associated inner product by $\langle\cdot,\cdot\rangle$.
Let a be a real continuous bilinear form on $\mathcal{H}\times\mathcal{H}$ which is <u>coercive</u>, i.e. there exists a constant $c > 0$ such that

$$a(u,u) \geq c\cdot\|u\|^2$$

for all $u \in \mathcal{H}$.
G. STAMPACCHIA [11] proved the following representation theorem:

(1.1) <u>THEOREM</u>: If L is a real continuous linear form on $\mathcal{H}$ and A a non-empty, closed, convex subset of $\mathcal{H}$ then there exists a uniquely determined $u \in A$ such that

$$a(u,v-u) \geq L(v-u)$$

for every $v \in A$. If w represents L, i.e. $\langle w,x\rangle = L(x)$

for every $x \in \mathcal{H}$, the map $w \longmapsto u$ is continuous.

NOTATIONS: 1) If in theorem (1.1) L is of the form $L(v) = a(w,v)$ for a certain $w \in \mathcal{H}$, we call the unique point $u \in A$ the a-projection of w and write $u = \pi_A^a(w)$. For the adjoint bilinear form $\hat{a}$ which is defined by

$$\hat{a}(u,v) := a(v,u) \qquad (u,v \in \mathcal{H})$$

we write $\hat{\pi}_A^a$ instead of $\pi_A^{\hat{a}}$.

2) Application of theorem (1.1) to the whole space yields the existence of two algebraic and topological automorphisms π^a and $\hat{\pi}^a$ of $\mathcal{H}$ which are defined by

$$a(\pi^a(v),w) = \langle v,w \rangle = a(v,\hat{\pi}^a(w)) \qquad (v,w \in \mathcal{H}).$$

3) In the special case that a is the inner product of $\mathcal{H}$, we shall omit the letter "a" in all definitions and notations.

The following approximation theorem of U.MOSCO can be found in [10]:

(1.2) THEOREM: Let $(A_i)_{i \in I}$ be an upward filtering (resp. downward filtering) family of non-empty, closed, convex subsets of $\mathcal{H}$. Then for any $w \in \mathcal{H}$, $\pi_{A_i}^a(w)$ converges strongly to $\pi_A^a(w)$ where $A = \overline{\bigcup_{i \in I} A_i}$ (resp. $A = \bigcap_{i \in I} A_i$ and $A \neq \emptyset$).

2. Potentials and copotentials

In the follwing, Ω will always denote a locally compact Hausdorff space and m a positive (Radon-) measure on Ω such that $m(\omega) > 0$ for all non-empty open subsets ω of Ω. By a function on Ω we always understand an extended real-valued m-measurable function on Ω. We do not distinguish notationally between a function and its class modulo locally m-negligible functions. We say that a property holds a.e.[m] in an m-measurable set $E \subset \Omega$ if the property holds in E except on a set which is locally of zero m-measure. For any open set ω of Ω we denote by $M_c(\omega)$ the set of m-essentially bounded functions on Ω with compact support contained in ω. The <u>support</u> S(f) of a function f on Ω is defined as the complement of the largest open set $\omega \subset \Omega$ such that $f = 0$ a.e.[m] on ω. Furthermore, for any open set $\omega \subset \Omega$ let $\mathcal{C}_c(\omega)$ be the space of continuous functions on Ω with compact support in ω provided with the usual topology. For any set $\mathcal{F}$ of functions, we define $\mathcal{F}^+ = \{f \in \mathcal{F} : f \geq 0 \text{ a.e.}[m]\}$, the positive functions of $\mathcal{F}$.

Following J.DENY [5], we give the definition of a functional space:

<u>DEFINITION</u>: A <u>functional space</u> (with respect to Ω and m) is a Hilbert space $\mathcal{H} = \mathcal{H}(\Omega, m)$ whose elements are (classes of) locally m-integrable functions on Ω such that the following axiom holds:

(F) For any compact subset K of Ω, there exists a constant $A(K) > 0$ such that

$$\int_K |u| dm \leq A(K) \cdot \|u\|$$

for any $u \in \mathcal{H}$.

In the following, let a be a fixed continuous coercive bilinear form on $\mathcal{H} \times \mathcal{H}$. An immediate consequence of the above definition and theorem (1.1) is the following

(2.1) <u>PROPOSITION</u> ([3]): For each $f \in M_c(\Omega)$, there exists a unique element U_f^a (resp. $\hat{U}_f^a$) in $\mathcal{H}$ such that for any $v \in \mathcal{H}$

$$a(U_f^a, v) = \int vf\, dm \quad (\text{resp. } a(v, \hat{U}_f^a) = \int vf\, dm).$$

The sets

$$\mathcal{H}_a = \{U_f^a : f \in M_c(\Omega)\} \quad \text{and} \quad \hat{\mathcal{H}}_a = \{\hat{U}_f^a : f \in M_c(\Omega)\}$$

are dense in $\mathcal{H}$.

DEFINITION: The element U_f^a (resp. $\hat{U}_f^a$) of (2.1) is called the a-potential (resp. a-copotential) generated by f The number

$$a(U_f^a, U_f^a) = a(U_f^a, U_f^a)$$

is the a-energy of f The elements of the closed convex cone

$$\mathcal{P}^a := \overline{\{U_f^a : f \in M_c^+(\Omega)\}}$$

are called pure a-potentials. More generally, we define for any open set ω of Ω

$$\mathcal{P}_\omega^a := \overline{\{U_f^a : f \in M_c^+(\omega)\}} .$$

Evidently, $\mathcal{P}_\omega^a$ is a closed convex subcone of $\mathcal{P}_\Omega^a = \mathcal{P}^a$, In an analogous manner, the closed convex cones $\hat{\mathcal{P}}_\omega^a$ are defined.

The next theorem gives an important characterization of pure a-potentials and more generally of the elements of $\mathcal{P}_\omega^a$:

(2.2) THEOREM ([3]): Let ω be an open subset of Ω. Then a necessary and sufficient condition for an element $u \in \mathcal{H}$ to be in $\mathcal{P}_\omega^a$ is

$$a(u,v) \geq 0 \quad \text{for all} \quad v \in \mathcal{H} \quad \text{such that} \quad v \geq 0 \quad \text{on } \omega.$$

Now we are interested in the question when all pure a-potentials are positive functions. Evidently, $\mathcal{P}^a \subset \mathcal{H}^+$ iff $\hat{\mathcal{P}}^a \subset \mathcal{H}^+$. The following proposition gives sufficient conditions of which (1) is also necessary in the case that a is the inner product of $\mathcal{H}$ (see N. ARONSZAJN - K. SMITH [1]).

(2.3) PROPOSITION: Each of the following conditions (1) and (2) is sufficient for $\mathcal{P}^a \subset \mathcal{H}^+$:

(1) For every element $u \in \mathcal{H}$, there exists an element $u' \in \mathcal{H}$ such that $u'(x) \geq |u(x)|$ a.e.[m] and $a(u + u', u - u') \geq 0$.

(2) For every element $u \in \mathcal{H}$, there exists an

element $u'' \in \mathcal{H}$ such that $u''(x) \geq u^+(x)$ a.e.[m] and $a(u + u'', u - u'') \geq 0$, where $u^+(x) = \sup(0,u(x))$.

Proof: Assume (1), let $u \in \mathcal{P}^a$ and u' the associated element of u.

Then: $c \cdot \|u'-u\|^2 \leq a(u'-u, u'-u) = a(u', u'-u) - a(u,u'-u)$
$\leq - 2a(u, u'-u) \leq 0$

by (2.2), hence $u = u' \in \mathcal{H}^+$.

The proof of (2) is analogous.

DEFINITION: A functional space $\mathcal{H} = \mathcal{H}(\Omega,m)$ is called regular, if $\mathcal{C}_c(\Omega) \cap \mathcal{H}$ is dense in $\mathcal{C}_c(\Omega)$ and in $\mathcal{H}$.

We assume from now on that $\mathcal{H}$ is regular.

Now we generalize the notion of a-potentials.

DEFINITION: An a-potential is an element $u \in \mathcal{H}$ such that there exists a (Radon-) measure μ on Ω satisfying

$$a(u,\varphi) = \int \varphi \, d\mu \quad \text{for each} \quad \varphi \in \mathcal{C}_c(\Omega) \cap \mathcal{H}.$$

If such a μ exists, it is unique and called the associated measure, and we write $u = U^a_\mu$. In the same manner, we define the a-copotentials $\hat{U}^a_\mu$.

(2.4) REMARK: For a measure μ on Ω, the a-potential U^a_μ exists iff the a-copotential $\hat{U}^a_\mu$ exists or iff the potential U_μ exists. Furthermore, we have the following equalities for all $v \in \mathcal{H}$:

$$a(U^a_\mu,v) = a(v,\hat{U}^a_\mu) = \langle U_\mu,v \rangle.$$

The following lemma will be essential in the sequel:

(2.5) LEMMA: Let f be a locally m-integrable function on Ω, and suppose that the a-potential U^a_f of f exists in $\mathcal{H}$. Then, there exists a family $(f_i)_{i \in I} \subset M^+_c(\Omega)$ and a family $(u_i)_{i \in I} \subset \mathcal{P}^a$ such that:

(1) $(f_i)_{i\in I}$ is upward filtering to f^+ ;

(2) $U^a_{f_i} - u_i \leq U^a_f$ for all $i \in I$;

(3) $(U^a_{f_i} - u_i)_{i\in I}$ converges strongly to U^a_f .

Proof: We choose a family $(f_i)_{i\in I} \subset M^+_c(\Omega)$ such that (1) holds. Define for each $i \in I$:

$$A_i: = U^a_{f_i} - \mathcal{P}^a .$$

Then $(A_i)_{i\in I}$ is an upward filtering system of closed convex subsets of $\mathcal{H}$ such that $(-\mathcal{P}^a) \subset A_i$ for all $i \in I$.

To show: $U^a_f \in A: = \overline{\bigcup_{i\in I} A_i}$.

For this purpose, we choose another family $(g_i)_{i\in I} \subset M^+_c(\Omega)$ which is upward filtering to f^- where $f^-(x): = \inf(0,f(x))$. For any $\varphi \in \mathcal{C}_c(\Omega) \cap \mathcal{H}$, we have

$$a(U^a_{f_i} - U^a_{g_i}, \varphi) = \int \varphi(f_i - g_i)dm \longrightarrow \int \varphi f\, dm = a(U^a_f, \varphi) ,$$

hence $(U^a_{f_i} - U^a_{g_i})_{i\in I}$ converges weakly to U^a_f because $\mathcal{C}_c(\Omega) \cap \mathcal{H}$ is dense in $\mathcal{H}$ and a is continuous. Since $U^a_{f_i} - U^a_{g_i} \in A_i$ for all $i \in I$, we have $U^a_f \in A$.

Now, we denote for every $i \in I$

$$v_i: = \pi^a_{A_i}(U^a_f) .$$

By theorem (1.2), $(v_i)_{i\in I}$ converges strongly to $\pi^a_A(U^a_f) = U^a_f$.

By the definition of the A_i, there exists for every $i \in I$ a pure a-potential u_i such that

$$v_i = U^a_{f_i} - u_i .$$

The lemma is proved by showing $v_i \leq U^a_f$ for all $i \in I$.

By the definition of v_i, we have for all $v \in A_i$

$$a(v - v_i, v_i) \geq a(v - v_i, U^a_f)$$

$$a(v_i - v, v_i) \leq a(v_i - v, U^a_f) .$$

Let $g \in M^+_c(\Omega)$ and take especially

$$v = U^a_{f_i} - (U^a_g + u_i) \in U^a_{f_i} - \mathcal{P}^a = A_i .$$

Since
$$v_i - v = U^a_g ,$$
we have
$$\int g\, v_i \, dm = a(U^a_g, v_i) = a(v_i - v, v_i) \leq a(v_i - v, U^a_f)$$
$$= a(U^a_g, U^a_f) = \int g\, U^a_f \, dm ,$$
hence
$$v_i \leq U^a_f \quad \text{a.e.}[m] .$$
⌟

3. Normalized contractions

DEFINITION: A normalized contraction T is a transformation of $\mathbb{R}$ into itself such that

(1) $|T(z') - T(z)| \leq |z'-z|$ for all $z, z' \in \mathbb{R}$;

(2) $T(0) = 0$.

EXAMPLES: 1) The modulus contraction: $T(z) := |z|$.

2) Every projection T of $\mathbb{R}$ onto a closed interval containing 0 is a normalized contraction.

3) The unit contraction T_I, defined by

$$T_I(z) := \begin{cases} 0, & z \leq 0 \quad ; \\ z, & 0 \leq z \leq 1 \; ; \\ 1, & z \geq 1 \; , \end{cases}$$

i.e. T_I is the projection of $\mathbb{R}$ onto the closed interval $I = [0,1]$.
Let $\mathcal{H} = \mathcal{H}(\Omega, m)$ be a functional space and a be a continuous coercive bilinear form on $\mathcal{H} \times \mathcal{H}$.

DEFINITION ([3],[9]): Let T be a normalized contraction. We say, T operates on $\mathcal{H}$ (with respect to a) if, for any $u \in \mathcal{H}$, the function $T \circ u \in \mathcal{H}$ and the inequality

$$a(u + T\circ u, u - T\circ u) \geq 0$$

holds.
If the unit contraction T_I operates on $\mathcal{H}$ with respect to a and $\overset{\vee}{a}$

then a is called a <u>Dirichlet form</u> on $\mathcal{H}$.

4. Resolvent and coresolvent

Let a be a continuous coercive bilinear form on a regular functional space $\mathcal{H} = \mathcal{H}(\Omega,m)$. Let $L^2 = L^2(\Omega,m)$ be the Hilbert space of real square m-integrable functions on Ω with the norm denoted by $\|\cdot\|_2$. Next, we shall prove the existence of a resolvent $(R_p^a)_{p>0}$ and coresolvent $(\hat{R}_p^a)_{p>0}$ on L^2 and $\mathcal{H}$.

(4.1) <u>THEOREM</u>: Let f be given in L^2 or in $\mathcal{H}$. For each real number $p > 0$ there exists a unique element $u \in A_f$ where

$$A_f = \{w \in \mathcal{H} : w - f \in L^2\}$$

such that for each $v \in L^2 \cap \mathcal{H}$

$$pa(u,v) + \int (u-f)v \, dm = 0.$$

<u>Proof</u>: We shall borrow an idea used by G. STAMPACCHIA in [11] and shall prove the existence of a unique $u \in A_f$ such that for all $v \in L^2 \cap \mathcal{H}$

$$pa(u,v) + \int (u-f)v \, dm + L(v) = 0 \qquad (+)$$

where L is a given continuous linear form on $\mathcal{H}$. Then the theorem follows by taking $L = 0$.

(i) Let α (resp. β) be the symmetric (resp. anti-symmetric) part of a, i.e.

$$\alpha(u,v) := \frac{a(u,v)+a(v,u)}{2}, \qquad \beta(u,v) := \frac{a(u,v)-a(v,u)}{2}$$

and let $B := \sup\{|\beta(u,v)| : u,v \in \mathcal{H} \text{ and } \|u\| = \|v\| = 1\} < +\infty$. We introduce new continuous bilinear forms a_t, $0 \leq t \leq 1$, on $\mathcal{H} \times \mathcal{H}$ by

$$a_t(u,v) = \alpha(u,v) + t\beta(u,v)$$

which are also coercive with the same constant c.

(ii) We show: If the above statement (+) is true for the bilinear form a_τ then it is also true for a_t where $t \leq \tau + t'$ and $t' < \frac{c}{B}$.

Indeed, let $y \in \mathcal{H}$ and

$$L_y(v) := L(v) + p(t-\tau)\beta(y,v) \quad (v \in \mathcal{H}).$$

Then L_y is a continuous linear form on $\mathcal{H}$. By assumption, there exists a unique element $F(y) \in A_f$ such that for all $v \in L^2 \cap \mathcal{H}$

$$pa_\tau(F(y),v) + \int (F(y) - f)v \, dm + L_y(v) = 0.$$

The map $F: \mathcal{H} \longrightarrow A_f$ is a contraction. If $y_1, y_2 \in \mathcal{H}$ then $F(y_2) - F(y_1) \in L^2 \cap \mathcal{H}$ and

$$\|F(y_2)-F(y_1)\|^2 \leq \frac{1}{c} \; a_\tau(F(y_2)-F(y_1), F(y_2)-F(y_1))$$

$$= \frac{1}{cp} [-\int (F(y_2)-F(y_1))^2 dm - p(t-\tau)\beta(y_2-y_1, F(y_2)-F(y_1))$$

$$\leq \frac{t'}{c} \; |\beta(y_2-y_1, F(y_2)-F(y_1))| \leq \frac{t'B}{c} \|y_2-y_1\| \cdot \|F(y_2)-f(y_1)\| ,$$

hence

$$\|F(y_2) - F(y_1)\| \leq \frac{t'B}{c} \|y_2-y_1\| .$$

But $\frac{t'B}{c} < 1$, i.e. F is a contraction of $\mathcal{H}$. This yields the existence of a unique $u \in \mathcal{H}$ with $u = F(u) \in A_f$, hence

$$pa_\tau(u,v) + \int (u-f)v \, dm + L_u(v) = 0$$

or

$$pa_t(u,v) + \int (u-f)v \, dm + L(v) = 0 \quad \text{for all} \quad v \in L^2 \cap \mathcal{H} .$$

(iii) The above statement is true for $a_o = \alpha$. This can be shown in the same way as lemma 3 of [2]. By (ii), the above statement holds for a_t with $t \leq t' < \frac{c}{B}$. Continuing the process, the desired result follows after a finite number of steps.

<u>DEFINITION</u>: Let f be given in L^2 or in $\mathcal{H}$ and let $p > 0$. We denote by $R_p^a f$ (resp $\hat{R}_p^a f$) the unique element of A_f such that by (4.1) for every $v \in L^2 \cap \mathcal{H}$

$$(*) \qquad pa(R_p^a f, v) + \int (R_p^a f - f)v \, dm = 0$$

(resp. $pa(v,\hat{R}^a_p f) + \int (\hat{R}^a_p f-f)v \, dm = 0$).

The operators

$$R^a_p, \hat{R}^a_p : \mathcal{H} \longrightarrow \mathcal{H} \quad \text{and} \quad L^2 \longrightarrow L^2$$

are called the a-resolvent and a-coresolvent operators.

From the characteristic property (*) of the a-resolvent, we get at once the following

(4.2) PROPOSITION: The operators R^a_p, $\hat{R}^a_p$ defined in $\mathcal{H}$ and in L^2 have the following properties:

(1) $pR^a_p f = U^a_{(f-R^a_p f)}$ $\quad (f \in \mathcal{H} \cup L^2)$

(2) $a(R^a_p f, R^a_p f) \leq a(f,f)$ $\quad (f \in \mathcal{H})$

(3) R^a_p is linear and bounded both in $\mathcal{H}$ and in L^2 with norms $\|R^a_p\| \leq \frac{\|a\|}{c}$ and $\|R^a_p\|_2 \leq 1$.

(4) $a(R^a_p f,g) = a(f, \hat{R}^a_p g)$ $\quad (f, g \in \mathcal{H})$

(5) $\lim_{p\to 0} R^a_p = I$ (identity operator), $\lim_{p\to\infty} R^a_p = 0$, strongly in $\mathcal{H}$ as well as in L^2.

(6) $pR^a_p - qR^a_q = (p-q)\, R^a_p R^a_q$ $\quad (p > 0,\ q > 0)$.

(7) $U^a_f = \lim_{p\to\infty} pR^a_p f$ $\quad (f \in M_c(\Omega))$.

(4.3) REMARKS: 1) If an element $u \in \mathcal{H}$ is the a-potential of the function f ($u = U^a_f$), we shall write

$$D^a u = f$$

and call f the a-Laclacian of u. The domain of the linear operator D^a is dense in $\mathcal{H}$ by (2.1).

2) For f given in $\mathcal{H}$, $R^a_p f$ is a solution of

$$u + p\, D^a u = f$$

by the characteristic property (*). This fact, together with the formula (6) of (4.2) shows that a suitable restriction of $(-D^a)$ is the infinitesimal generator of a uniquely determined equi-continuous semi-group of bounded linear operators on $\mathcal{H}$ of classes (C_o) (see K. YOSIDA [13], p. 248).

DEFINITION: The a-resolvent $(R_p^a)_{p>0}$ is said to be positive (resp. sub-markovian) if for every $p > 0$, the a-resolvent operator R_p^a is positive (resp. sub-markovian), i.e.

$$R_p^a f \geq 0 \quad (\text{resp. } 0 \leq R_p^a f \leq 1)$$

for every $f \in \mathcal{C}_c(\Omega)$ such that $f \geq 0$ (resp. $0 \leq f \leq 1$).

(4.4) LEMMA: Let T be a normalized contraction which operates on $\mathcal{H}$ with respect to a. If T leaves $f \in \mathcal{C}_c(\Omega)$ invariant, the same holds for $R_p^a f$.

Proof: Let $f \in \mathcal{C}_c(\Omega)$ such that $T \circ f = f$. Since

$$|T\circ R_p^a f - R_p^a f| \leq |T\circ R_p^a f - f| + |R_p^a f - f|$$
$$\leq |T\circ R_p^a f - T\circ f| + |R_p^a f - f| \leq 2|R_p^a f - f|$$

and $R_p^a f - f \in L^2$, we have

$$T\circ R_p^a f - R_p^a f \in \mathcal{H} \cap L^2 .$$

Considering that T operates on $\mathcal{H}$ with respect to a, we get the following estimates:

$$0 \leq \frac{pc}{2} \|R_p^a f - T\circ R_p^a f\|^2 \leq \frac{p}{2}\, a(R_p^a f - T\circ R_p^a f,\ R_p^a f - T\circ R_p^a f)$$
$$= \frac{p}{2}\, a(R_p^a f,\ R_p^a f - T\circ R_p^a f) - \frac{p}{2}\, a(T\circ R_p^a f,\ R_p^a f - T\circ R_p^a f)$$
$$\leq pa(R_p^a f,\ R_p^a f - T\circ R_p^a f)$$
$$= \int (f - R_p^a f)\,(R_p^a f - T\circ R_p^a f)\,dm$$
$$= \int (f - R_p^a f)\,(R_p^a f - f + T\circ f - T\circ R_p^a f)\,dm$$
$$= -\int (f - R_p^a f)^2 dm + \int (f - R_p^a f)\,(T\circ f - T\circ R_p^a f)\,dm$$
$$\leq -\int (f - R_p^a f)^2 dm + \int |f - R_p^a f|\,|T\circ f - T\circ R_p^a f|\,dm$$
$$\leq -\int (f - R_p^a f)^2 dm + \int |f - R_p^a f|^2 dm = 0 ,$$

hence $T \circ R_p^a f = R_p^a f$. ∎

(4.5) <u>COROLLARY</u>: If the modulus contraction (resp. unit contraction) operates on $\mathcal{H}$ with respect to a then the a-resolvent $(R_p^a)_{p>0}$ is positive (resp. sub-markovian).

(4.6) <u>PROPOSITION</u>: If the a-resolvent operator R_p^a is positive, then there exists a unique measure σ_p^a on the product space $\Omega \times \Omega$ such that

$$\int R_p^a f(x) g(x) m(dx) = \iint f(x) g(y) \sigma_p^a(dx\, , dy)$$

for every f, $g \in \mathcal{C}_c(\Omega)$.

<u>Proof</u>: The map $(f,g) \longmapsto \int R_p^a f \cdot g \, dm$ defines a positive linear functional on $\mathcal{C}_c^+(\Omega) \times \mathcal{C}_c^+(\Omega)$ which is a total class in $\mathcal{C}_c(\Omega \times \Omega)$. Hence, there exists a unique positive measure σ_p^a on $\Omega \times \Omega$ which satisfies the desired equality. __/

(4.7) <u>PROPOSITION</u>: If the a-resolvent operator R_p^a is sub-markovian then the measure σ_p^a of (4.6) satisfies the inequality

$$\sigma_p^a(\Omega \times K) \leq m(K)$$

for any compact set $K \subset \Omega$.

<u>Proof</u>: By standard approximation, (4.6) also holds for $g = \chi_K$ where $\chi_K(x) = \begin{cases} 1, & x \in K \\ 0, & x \notin K \end{cases}$.

Let $f \in \mathcal{C}_c(\Omega)$ with $0 \leq f \leq 1$. Then $0 \leq R_p^a f \leq 1$, whence

$$\iint f(x) \chi_K(y) \sigma_p^a(dx,dy) = \int_K R_p^a f \, dm \leq m(K).$$

Thus

$$\sigma_p^a(\Omega \times K) = \sup\{\iint f(x)\chi_K(y)\sigma_p^a(dx,dy) : f \in \mathcal{C}_c(\Omega),\ 0 \leq f \leq 1\} \leq m(K). \quad \_\_/$$

(4.8) <u>REMARK</u>: By the preceding proposition (4.7), there exists for a sub-markovian a-resolvent operator R_p^a a locally m-integrable function m_p^a on Ω such that for all $g \in \mathcal{C}_c(\Omega)$

$$\iint g(y) \sigma_p^a(dx,dy) = \int g(y) m_p^a(y) m(dy),$$

Evidently, $0 \leq m_p^a \leq 1$.

In the same manner, the measure δ_p^a (resp. function $\hat{m}_p^a$) is defined for a positive (resp. sub-markovian) a-coresolvent operator $\hat{R}_p^a$).

5. The associated approximation form

Let a be a continuous coercive bilinear form on a regular functional space $\mathcal{H} = \mathcal{H}(\Omega,m)$.

DEFINITION: The approximation form H_p^a of index $p > 0$ is defined for every $f, g \in \mathcal{C}_c(\Omega)$ by:

$$H_p^a(f,g) := \frac{1}{p} \int (f-R_p^a f)g \, dm \quad .$$

(5.1) LEMMA: For every $f, g \in \mathcal{C}_c(\Omega) \cap \mathcal{H}$, we have

$$\lim_{p \to 0} H_p^a(f,g) = a(f,g)$$

Proof: Using property (5) of (4.2), we get

$$\lim_{p\to 0} H_p^a(f,g) = \lim_{p\to 0} \frac{1}{p} \int (f-R_p^a f)g \, dm = \lim_{p\to 0} a(R_p^a f,g) = a(f,g). \quad \_/$$

(5.2) PROPOSITION: A necessary and sufficient condition for $f \in \mathcal{C}_c(\Omega)$ to be in $\mathcal{H}$ is that $H_p^a(f,f)$ is bounded for p.

Proof: The necessity of the above assertion follows from (5.1). So assume $H_p^a(f,f) \leq B$ for every $p > 0$. The inequality $\int (f - R_p^a f)^2 dm \geq 0$ implies

$$\int (f - R_p^a f) f \, dm \geq \int (f - R_p^a f) R_p^a f \, dm$$

which yields

$$c \cdot \|R_p^a f\|^2 \leq a(R_p^a f, R_p^a f) = \frac{1}{p} \int (f - R_p^a f) R_p^a f \, dm$$
$$\leq \frac{1}{p} \int (f - R_p^a f) f \, dm = H_p^a(f,f) \leq B \ ,$$

hence $\|R_p^a f\|$ is bounded with respect to p. Therefore, we may assume the existence of an element $u \in \mathcal{H}$ such that

$$\lim_{p\to 0} R_p^a f = u \quad \text{weakly in } \mathcal{H}$$

which implies $R_p^a f(x) \longrightarrow u(x)$ a.e.[m] on Ω .

On the other hand, we have by property (5) of (4.2)

$$\lim_{p\to o} R_p^a f = f \quad \text{strongly in } L^2$$

which implies

$$R_p^a f(x) \longrightarrow f(x) \quad \text{as} \quad p \longrightarrow 0 \quad \text{a.e.}[m] \text{ on } \Omega ,$$

consequently $u = f$ a.e.[m] om Ω .
Thus $f = u \in \mathcal{H}$. ⌟

(5.3) <u>REMARKS</u>: 1) If the a-resolvent $(R_p^a)_{p>o}$ is positive, we have for every $f, g \in \mathcal{C}_c(\Omega)$ by (4.6)

$$H_p^a(f,g) = \frac{1}{p} [\int fg \, dm - \iint f(x)g(y)\sigma_p^a(dx,dy)]$$

2) If $(R_p^a)_{p>o}$ and $(\hat{R}_p^a)_{p>o}$ are sub-markovian, we have for every $f, g \in \mathcal{C}_c(\Omega)$ by (4.8)

(i) $H_p^a(f,g) = \frac{1}{p} [\int(1-m_p^a)fg \, dm + \iint f(x)[g(x)-g(y)]\sigma_p^a(dx,dy)]$

(ii) $H_p^a(f,f) = \frac{1}{2p}[\int(2-m_p^a-\hat{m}_p^a)f^2 dm + \iint |f(x)-f(y)|^2\sigma_p^a(dx,dy)]$.

6. <u>The domination principle</u>

Let a be a continuous coercive bilinear form on a regular functional space $\mathcal{H} = \mathcal{H}(\Omega,m)$.

<u>DEFINITION</u>: We say that $\mathcal{P}^a$ satisfies the <u>domination principle</u> if for any $f \in M_c^+(\Omega)$ and any $u \in \mathcal{P}^a$ the condition

$$U_f^a(x) \leq u(x)$$

holds a.e.[m] on Ω if it holds a.e.[m] on the set $\{x \in \Omega : f(x) > 0\}$.

<u>DEFINITION</u>: We say that $\mathcal{P}^a$ satisfies the <u>principle of the lower envelope</u> if $\inf(u,v) \in \mathcal{P}^a$ for any two $u, v \in \mathcal{P}^a$.

A first relation between normalized contractions, resolvents and the above defined principles is given in the next theorem.

(6.1) <u>THEOREM</u>: The following ten conditions are equivalent:

(1) The modulus contraction operates on $\mathcal{H}$ with respect to a.

(2) $\mathcal{P}^a \subset \mathcal{H}^+$, and $\mathcal{P}^a$ satisfies the principle of the lower envelope.

(3) $\mathcal{P}^a$ satisfies the principle of the lower envelope.

(4) $\mathcal{P}^a$ satisfies the domination principle.

(5) The a-resolvent $(R_p^a)_{p>o}$ is positive.

(6) The modulus contraction operates on $\mathcal{H}$ with respect to $\hat{a}$.

(7) $\hat{\mathcal{P}}^a \subset \mathcal{H}^+$, and $\hat{\mathcal{P}}^a$ satisfies the principle of the lower envelope.

(8) $\hat{\mathcal{P}}^a$ satisfies the principle of the lower envelope.

(9) $\hat{\mathcal{P}}^a$ satisfiee the domination principle.

(10) The a-coresolvent $(\hat{R}_p^a)_{p>o}$ is positive.

<u>Proof</u>: (1) $\Longleftrightarrow$ (6): $a(u+|u|, u-|u|) \geq 0$ implies $a(u-|u|, u+|u|) \geq 0$ by replacing u by $-u$.
We remark that the addition of both inequalities yields

$$a(|u|,|u|) \leq a(u,u), \quad \text{hence} \quad \| |u| \| \leq \sqrt{\frac{\|a\|}{c}} \cdot \|u\| .$$

(1) $\Rightarrow$ (2): $\mathcal{P}^a \subset \mathcal{H}^+$ by (2.3).
To show: $u, v \in \mathcal{P}^a$ implies $\inf(u,v) \in \mathcal{P}^a$.
Define a non-empty, closed, convex set A by

$$A: = \{w \in \mathcal{H} : w \geq \inf(u,v)\}$$

and let

$$f = \pi_A^a(0) ,$$

i.e. f is the unique element of A such that

$$a(f,w-f) \geq 0 \quad \text{for all } w \in A \text{ (theorem (1.1))}.$$

If $w' \in \mathcal{H}^+$, then $w = w' + f \in A$, hence

$$a(f,w') = a(f,w-f) \geq 0 ,$$

thus $f \in \mathcal{P}^a$ by (2.2).
We shall show: $f = \inf(u,v)$.
If $g := \inf(u,f) \in A$, then $a(f,g-f) \geq 0$. But we have also $a(g,f-g) \geq 0$.

Indeed, by the definition of g:

$$g = \inf(u,f) = \frac{u+f}{2} - \frac{|u-f|}{2} \ ,$$

we have the following estimates:

$$\begin{aligned} 4a(g,f-g) &= a(u+f - |u-f|, \ |u-f| - (u-f)) \\ &= a(u+f, \ |u-f| - (u-f)) + a(|u-f|, \ (u-f) - |u-f|) \\ &\geq a(u+f, \ |u-f| - (u-f)) + a(u-f, \ |u-f| - (u-f)) \\ &= 2a(u, \ |u-f| - (u-f)) \geq 0 \ , \end{aligned}$$

where (2.2) yields the latter inequality. Hence $a(g, f-g) \geq 0$. Therefore,

$$0 < c \cdot \|g-f\|^2 \leq a(g-f,g-f) = a(g,g-f) - a(f,g-f) \leq 0$$

which implies $f = g = \inf(u,f)$ and finally $f \leq u$.
In the same way, we get $f \leq v$, thus $f \leq \inf(u,v) \leq f$, i.e. $\inf(u,v) = f \in \mathcal{S}^a$.

(2) $\Longrightarrow$ (3): trivial.

(3) $\Longrightarrow$ (4): Given $f \in M_c^+(\Omega)$, $u \in \mathcal{S}^a$, and assume

$$U_f^a(x) \leq u(x) \quad \text{a.e.}[m] \text{ on } \{x \in \Omega\colon f(x) > 0\}.$$

By assumption, $v := \inf(U_f^a,u) \in \mathcal{S}^a$, hence by (2.2)

$$a(v, \ U_f^a - v) \geq 0 \ .$$

On the other hand,

$$a(U_f^a, \ U_f^a - v) = \int (U_f^a - v)f \, dm = 0,$$

since $U_f^a(x) = v(x)$ a.e.[m] on $\{x \in \Omega : f(x) > 0\}$.
Thus we get

$$0 \leq c \cdot \|U_f^a - v\|^2 \leq a(U_f^a - v, \ U_f^a - v) = a(U_f^a, \ U_f^a - v) - a(v, U_f^a - v) \leq 0$$

which implies

$$U_f^a = v = \inf(U_f^a,u) \leq u \ .$$

(4) $\Longrightarrow$ (5): Let $g \in \mathcal{C}_c^+(\Omega)$.

To show: $R_p^a g \geq 0$.

By property (1) of (4.2), $pR_p^a g$ is the a-potential generated by $g - R_p^a g$. Since the pure a-potential U_g^a exists in $\mathcal{H}$, there exists an a-potential $U^a_{R_p^a g}$ in $\mathcal{H}$. Therefore we have

$$(*) \qquad pR_p^a g = U^a_{g-R_p^a g} = U_g^a - U^a_{R_p^a g}$$

or

$$(**) \qquad U_g^a - pR_p^a g = U^a_{R_p^a g} .$$

Now choose for $f: = R_p^a g$ a family $(f_i)_{i\in I} \subset M_c^+(\Omega)$ and a family $(u_i)_{i\in I} \subset \mathcal{S}^a$ such that the conditions of (2.5) are satisfied, i.e.

$$(f_i)_{i\in I} \text{ increases to } f, \quad U^a_{f_i} - u_i \leq U_f^a, \text{ and}$$

$$\lim (U^a_{f_i} - u_i) = U_f^a \text{ strongly in } \mathcal{H} .$$

Then (**) implies

$$(***) \qquad U^a_{f_i} - u_i \leq U_f^a = U_g^a - pf \quad \text{for all } i \in I .$$

Define for all $i \in I$:

$$Y_i: = \{y \in \Omega : f_i(y) > 0\} ,$$

and

$$Y : = \{y \in \Omega : f(y) > 0\} .$$

By the definition of f_i, we have $Y_i \subset Y$ for all $i \in I$.

The inequality (***) implies a.e.[m] on Y_i:

$$U^a_{f_i} - u_i \leq U_g^a ,$$

hence by the domination principle

$$U^a_{f_i} - u_i \leq U_g^a \text{ a.e.[m] on } \Omega \text{ for all } i \in I,$$

because $U_g^a + u_i \in \mathcal{S}^a$.

Since strong convergence in $\mathcal{H}$ implies convergence a.e.[m] on Ω, the latter inequality yields

$$U_f^a \leq U_g^a \quad \text{a.e.}[m] \quad \text{on} \quad \Omega ,$$

and therefore by (*)

$$pR_p^a f = U_g^a - U_f^a \geq 0 \quad \text{a.e.}[m] \quad \text{on } \Omega ,$$

and (5) is proved.

(5) $\Longrightarrow$ (1): We have to show that $u \in \mathcal{H}$ implies $|u| \in \mathcal{H}$ and

$$a(u + |u|, u - |u|) \geq 0.$$

First, let $f \in \mathcal{C}_c(\Omega) \cap \mathcal{H}$, then $|f| \in \mathcal{C}_c(\Omega)$ and by the following estimates

$$H_p^a(|f|,|f|) = \frac{1}{p}[\int |f|^2 dm - \iint |f(x)|\cdot|f(y)|\sigma_p^a(dx,dy)]$$

$$\leq \frac{1}{p}[\int f^2\, dm - \iint f(x)f(y)\sigma_p^a(dx,dy)] = H_p^a(f,f) ,$$

we have $|f| \in \mathcal{C}_c(\Omega) \cap \mathcal{H}$ according to (5.2) and

$$a(|f|, |f|) \leq a(f,f)$$

by (5.1). Since

$$H_p^a(f+|f|,f-|f|)$$

$$= \frac{1}{p}[\int (f+|f|)(f-|f|)dm - \iint [f(x)+|f(x)|][f(y)-|f(y)|]\sigma_p^a(dx,dy)]$$

$$= \frac{1}{p}\iint [f(x)+|f(x)|][|f(y)|-f(y)]\sigma_p^a(dx,dy) \geq 0 ,$$

the desired inequality $a(f + |f|, f - |f|) \geq 0$ follows again by (5.1).

For any $u \in \mathcal{H}$, there exists a sequence $(f_n) \subset \mathcal{C}_c(\Omega) \cap \mathcal{H}$ converging strongly to u in $\mathcal{H}$. By the foregoing results, we have

$$|f_n| \in \mathcal{C}_c(\Omega) \cap \mathcal{H},\ a(|f_n|,|f_n|) \leq a(f_n,f_n),\ a(f_n+|f_n|,f_n-|f_n|) \geq 0 .$$

Therefore $(\|f_n\|)$ is bounded, and we may assume the existence of an element $u_o \in \mathcal{H}$ such that $|f_n|$ converges weakly to u_o, which implies the convergence $|f_n| \longrightarrow u_o$ a.e.[m] on Ω. But by definition, $|f_n| \longrightarrow |u|$ a.e.[m] on Ω, hence $|u| = u_o \in \mathcal{H}$. To show the desired inequality, we remark that the map

$$v \longmapsto a(v,v)$$

is lower semi-continuous with respect to the weak topology in $\mathcal{H}$. Using this, we get

$$0 \leq \limsup_{n \to \infty} a(f_n+|f_n|,f_n-|f_n|)$$

$$\leq \limsup_{n \to \infty} a(f_n,f_n)+\limsup_{n \to \infty} [-a(|f_n|,|f_n|)]+\limsup_{n \to \infty} a(|f_n|,f_n)$$

$$+\limsup_{n \to \infty} [-a(f_n,|f_n|)]$$

$$\leq a(u,u)-a(|u|,|u|)+a(|u|,u)-a(u,|u|) = a(u+|u|,\ u-|u|)$$

where the relations

$$\limsup_{n \to \infty} a(|f_n|,f_n) = a(|u|,u) \text{ and } \limsup_{n \to \infty} [-a(f_n,|f_n|)] = -a(u,|u|)$$

follow at once from

(6.2) <u>LEMMA</u>: Let (f_n), $(g_n) \subset \mathcal{H}$ such that f_n converges strongly to $f \in \mathcal{H}$ and g_n weakly to $g \in \mathcal{H}$ in $\mathcal{H}$. Then

$$\lim_{n\to\infty} a(f_n,g_n) = a(f,g)$$

<u>Proof</u>: According to the equality

$$a(f_n,g_n) - a(f,g) = a(f_n-f,g_n) + a(f,g_n-g) ,$$

we get the assertion. ⏌

For the rest of this section, we assume that the modulus contraction operates on $\mathcal{H}$ with respect to a.

(6.3) <u>PROPOSITION</u>:

1) $\mathscr{C}^+_c(\Omega) \cap \mathcal{H}$ is a total class in $\mathscr{C}_c(\Omega)$.

2) $\mathscr{C}^+_c(\Omega) \cap \mathcal{H}$ is dense in $\mathcal{H}^+$.

<u>Proof</u>: 1) Let $f \in {}^+_c(\Omega)$, $\epsilon > 0$ and a neighborhood V of the support S(f) of f be given. Since $\mathscr{C}_c(\Omega) \cap \mathcal{H}$ is dense in $\mathscr{C}_c(\Omega)$, there exists a $\varphi \in \mathscr{C}_c(\Omega) \cap \mathcal{H}$ such that $S(\varphi) \subset V$ and

$$\sup_{x\in\Omega} |f(x) - \varphi(x)| < \varepsilon .$$

But $|\varphi| \in \mathscr{C}_c^+(\Omega) \cap \mathscr{H}$, $S(|\varphi|) \subset V$, and

$$\sup_{x\in\Omega} |f(x) - |\varphi(x)|| \leq \sup_{x\in\Omega} |f(x) - \varphi(x)| < \varepsilon .$$

2) Let $u \in \mathscr{H}^+$, and $(u_n) \subset \mathscr{C}_c(\Omega) \cap \mathscr{H}$ such that u_n converges strongly in $\mathscr{H}$ to u.

Then $|u_n| \in \mathscr{C}_c^+(\Omega) \cap \mathscr{H}$, and furthermore, $|u_n|$ converges weakly in $\mathscr{H}$ to u. In fact, we have for all $\hat{U}_f^a \in \hat{\mathscr{H}}_a$ first of all

$$|a(|u_n|-u, \hat{U}_f^a)| = |\int [|u_n|-u] f\, dm| \leq \int ||u_n|-f|\cdot|f|dm$$
$$\leq \int |u_n-u|\cdot|f|dm = a(|u_n-u|\ \hat{U}^a_{|f|})$$
$$\leq \|a\|\cdot\||u_n-u|\|\cdot\|\hat{U}^a_{|f|}\| \leq \|a\|\cdot\sqrt{\frac{\|a\|}{c}}\cdot\|u_n-u\|\cdot\|\hat{U}^a_{|f|}\| ,$$

and secondly

$$\limsup_{n\to\infty} \||u_n|\| \leq \sqrt{\frac{\|a\|}{c}} \limsup_{n\to\infty} \|u_n\| = \sqrt{\frac{\|a\|}{c}} \|u\| ,$$

and this means weak convergence of $|u_n|$ to u, since $\hat{\mathscr{H}}_a$ is dense in $\mathscr{H}$ by (2.1). Hence $\mathscr{H}^+$ is the weak closure of $\mathscr{C}_c^+(\Omega) \cap \mathscr{H}$, therefore it is also the strong closure, because $\mathscr{C}_c^+(\Omega) \cap \mathscr{H}$ is a convex cone.

(6.4) <u>COROLLARY</u>: Let $f \in \mathscr{H} \cup L^2$ with $f \geq 0$. Then $R_f^a \geq 0$.

<u>Proof</u>: We have $R_p^a\varphi \geq 0$ for all $\varphi \in \mathscr{C}_c^+(\Omega)$. By property (3) of (4.2), R_p^a is continuous on $\mathscr{H}$ and on L^2. This fact together with (6.3) imply the assertion.

(6.5) <u>COROLLARY</u>: $\mathscr{P}^a = \{U_\mu^a : \mu$ positive measure on $\Omega\}$.

<u>Proof:</u> For $u \in \mathscr{P}^a$,

$$\varphi \longmapsto a(u,\varphi)$$

defines by (2.2) a positive linear functional on $\mathscr{C}_c^+(\Omega) \cap \mathscr{H}$. According to (6.3), there exists a unique positive measure μ on Ω such that

$$a(u,\varphi) = \int \varphi \, d\mu$$

for all $\varphi \in \mathcal{C}_c(\Omega) \cap \mathcal{H}$, i.e. $u = U^a_\mu$.

Conversely, for $\varphi \in \mathcal{C}^+_c(\Omega) \cap \mathcal{H}$

$$a(U^a_\mu,\varphi) = \int \varphi \, d\mu \geq 0.$$

If $v \in \mathcal{H}^+$, then by (6.3) there exists a sequence $(\varphi_n) \subset \mathcal{C}^+_c(\Omega) \cap \mathcal{H}$ which converges strongly to v in $\mathcal{H}$, hence

$$a(U^a_\mu, v) = \lim_{n\to\infty} a(U^a_\mu,\varphi_n) \geq 0 ,$$

thus $U^a_\mu \in \mathcal{P}^a$ by (2.2). _/

7. The associated kernel and singular measure

In this section, we assume that the modulus contraction operates on a regular functional space $\mathcal{H} = \mathcal{H}(\Omega,m)$ with respect to a continuous coercive bilinear form a.

(7.1) LEMMA: There exists a uniquely determined positive measure $\varkappa^a$ on the product space, called the kernel of a, such that for all $f, g \in M_c(\Omega)$

$$a(U^a_f, \hat{U}^a_g) = \iint f(x)g(y)\, \varkappa^a(dx,dy) .$$

Proof: The map

$$(f,g) \longmapsto a(U^a_f, \hat{U}^a_g) = \int U^a_f \cdot g \, dm$$

defines a positive linear functional on $\mathcal{C}^+_c(\Omega) \times \mathcal{C}^+_c(\Omega)$ which implies the assertion. _/

The kernel $\varkappa^a$ can be obtained also in the following way:

(7.2) PROPOSITION: $\varkappa^a = \lim_{p\to\infty} p\sigma^a_p$ (in the vague topology).

Proof: Since the a-resolvent $(R^a_p)_{p>0}$ is positive, $p\sigma^a_p$ is increasing with respect to p by property (6) of (4.2).

According to

$$p \iint f(x)g(y)\,\sigma_p^a(dx,dy) = p \int R_p^a f \cdot g \, dm$$
$$= a(pR_p^a f, \hat{U}_g^a) \leq \|a\| \cdot \|pR_p^a f\| \cdot \|\hat{U}_g^a\|$$
$$\leq \|a\| \cdot \|U_f^a\| \cdot \|\hat{U}_g^a\|$$

by property (7) of (4.2) for all $f, g \in \mathcal{C}_c^+(\Omega)$, $(p\sigma_p^a)_{p>0}$ is bounded, hence converges vaguely [as p tends to infinity] to some positive measure on $\Omega \times \Omega$ which has to be the kernel $\varkappa^a$ by property (7) of (4.2).

(7.3) REMARKS: 1) For $f \in M_c(\Omega)$, we have by (7.1)

$$0 \leq a(U_f^a, U_f^a) = a(U_f^a, \hat{U}_f^a) = \iint f(x)f(y)\,\varkappa^a(dx,dy),$$

hence $\varkappa^a$ is a measure of "positive type".

2) Given $f \in M_c(\Omega)$, the a-potential U_f^a is by (7.1) the density of the projection of the measure $f(x)\,\varkappa^a(dx,dy)$ on its second factor.

Let Δ be the diagonal set of $\Omega \times \Omega$, i.e.

$$\Delta = \{(x,x) \in \Omega \times \Omega : x \in \Omega\} .$$

By (6.3), the set

$$\mathfrak{F}_\Delta = \{f \otimes g : f,g \in \mathcal{C}_c^+(\Omega) \cap \mathcal{H} \;,\; S(f) \cap S(g) = \emptyset\}$$

is a total class in $\mathcal{C}_c(\Omega \times \Omega \setminus \Delta)$ where

$$f \otimes g(x,y) = f(x)g(y) \qquad [(x,y) \in \Omega \times \Omega].$$

(7.4) LEMMA: There exists a unique positive measure σ^a on $\Omega \times \Omega \setminus \Delta$, called the singular measure of a, such that for all $f \otimes g \in \mathfrak{F}_\Delta$

$$a(f,g) = -2 \iint f(x)g(y)\,\sigma^a(dx,dy)$$

Proof: It suffices to show $a(f,g) \leq 0$ for all $f \otimes g \in \mathfrak{F}_\Delta$. Let $u = f - g$, then $|u| = f + g$, hence

$$0 \leq a(u+|u|,\ u-|u|) = -4a(f,g).$$ ⌟

The relation between the measures σ^a and σ^a_p is given in:

(7.5) PROPOSITION: $\sigma^a = \lim_{p\to o} \frac{1}{2p}\ \sigma^a_p$ on $\Omega \times \Omega \setminus \Delta$ (in the vague topology).

Proof: Let $f \otimes g \in \mathcal{F}_\Delta$, then, using property (5) of (4.2), we obtain the desired relation:

$$\lim_{p\to o} \frac{1}{2p} \iint f(x)g(y)\,\sigma^a_p(dx,dy) = \lim_{p\to o} \frac{1}{2p} \int R^a_p f \cdot g\ dm$$

$$= - \lim_{p\to o} \frac{1}{2p} \int (f-R^a_p f) g\ dm = - \lim_{p\to o} \frac{1}{2}\ a(R^a_p f, g)$$

$$= - \frac{1}{2}\ a(f,g) = \iint f(x)g(y)\,\sigma^a(dx,dy)\ .$$

8. The complete maximum principle

Let a again be a continuous coercive bilinear form on a regular functional space $\mathcal{H} = \mathcal{H}(\Omega,m)$. The aim of this section is to characterize those a which are Dirichlet forms. We start with two definitions.

DEFINITION: We say, that $\mathcal{P}^a$ satisfies the complete maximum principle if for any $f \in M^+_c(\Omega)$ and any $u \in \mathcal{P}^a$, the condition

$$U^a_f \leq u + 1$$

holds a.e.[m] on Ω, provided it holds a.e.[m] on the set $\{x \in \Omega : f(x) > 0\}$.

DEFINITION: We say, that $\mathcal{P}^a$ satisfies the strong principle of the lower envelope if $\inf(u,v+1) \in \mathcal{P}^a$ for any two $u,v \in \mathcal{P}^a$.

(8.1) THEOREM: The following four conditions are equivalent:

(1) a is a Dirichlet form on $\mathcal{H}$.

(2) $\mathcal{P}^a$ and $\hat{\mathcal{P}}^a$ satisfy the strong principle of the lower envelope.

(3) $\mathcal{P}^a$ and $\hat{\mathcal{P}}^a$ satisfy the complete maximum principle.

(4) The a-resolvent $(R_p^a)_{p>o}$ and a-coresolvent $(\hat{R}_p^a)_{p>o}$ are sub-markovian.

<u>Proof</u>: (1) $\Longrightarrow$ (2): By (4.5), $(R_p^a)_{p>o}$ is sub-markovian and therefore positive, hence $\mathcal{P}^a$ satisfies by (6.1) the principle of the lower envelope. Because of the equation

$$\inf(u,v+1) = \inf(u,v+\inf(u,1)),$$

it suffices to show $\inf(u,1) \in \mathcal{P}^a$ for every $u \in \mathcal{P}^a$. Define $f: = \pi_A^a(0)$ where the non-empty, closed, convex set A is defined by $A: = \{w \in \mathcal{H} : w \geq \inf(u,1)\}$.
By the definition of f, we have for all $w \in A$

$$a(f,w-f) \geq 0 .$$

If $w' \in \mathcal{H}^+$, then $w = w' + f \in A$, hence

$$a(f,w') = a(f,w-f) \geq 0,$$

thus $f \in \mathcal{P}^a$ by (2.2).
We shall show: $f = \inf(u,1)$.
In the same way as in the proof of ((1)$\Longrightarrow$(2)) of (6.1), we obtain

$$f = \inf(u,f), \quad \text{i.e.} \quad f \leq u .$$

Next, let $h = \inf(1,f)$, hence $h \in A$, and therefore

$$a(f,h-f) \geq 0 .$$

Since T_I operates on $\mathcal{H}$ with respect to a and since $h = T_I \circ f$, we have

$$a(f+h, f-h) \geq 0.$$

The last two inequalities yield the following estimates:

$$0 \leq c\cdot\|f-h\|^2 \leq a(f-h,f-h) = a(f,f-h) - a(h,f-h)$$

$$\leq -2a(f,h-f) \leq 0 ,$$

hence $f = h = \inf(1,f)$ or $f \leq 1$.

Finally we get

$$f \leq \inf(u,1) \leq f,$$

i.e. $\inf(u,1) = f \in \mathcal{P}^a$.

(2) $\Rightarrow$ (3): Given $f \in M_c^+(\Omega)$, $u \in \mathcal{P}^a$, and assume

$$U_f^a(x) \leq u(x) + 1 \quad \text{a.e.}[m] \text{ on } \{x \in \Omega: f(x) > 0\},$$

By assumption, $v = \inf(U_f^a, u+1) \in \mathcal{P}^a$, hence by (2.2)

$$a(v, U_f^a-v) \geq 0.$$

On the other hand,

$$a(U_f^a, U_f^a-v) = \int (U_f^a-v)f\,dm = 0 ,$$

since $U_f^a(x) = v(x)$ a.e.[m] on $\{x \in \Omega : f(x) > 0\}$.

Thus we get

$$0 \leq c\|U_f^a-v\|^2 \leq a(U_f^a-v, U_f^a-v)$$

$$= a(U_f^a, U_f^a-v) - a(v, U_f^a-v) \leq 0$$

which implies

$$U_f^a = v = \inf(U_f^a, u+1) \leq u+1 .$$

(3) $\Rightarrow$ (4): Let $g \in \mathcal{C}_c(\Omega)$ such that $0 \leq g \leq 1$.

To show: $0 \leq R_p^a g \leq 1$.

Obviously, the complete maximum principle implies the domination principle, hence we have $R_p^a g \geq 0$ by (6.1). By property (1) of (4.2), $pR_p^a g$ is the a-potential generated by $g - R_p^a g$. Writing $f = R_p^a g$, we have, therefore,

(*) $$U_{g-f}^a = pf.$$

Now choose families $(f_i)_{i\in I} \in M_c^+(\Omega)$, $(u_i)_{i\in I} \in \mathcal{P}^a$ such that by (2.5):

$(f_i)_{i\in I}$ is upward filtering to $(g-f)^+$,

$U^a_{f_i}-u_i \leq U^a_{g-f}$, and $(U^a_{f_i}-u_i)_{i\in I}$ converges strongly to U^a_{g-f}.

The equation (*) implies

$$U^a_{f_i}-u_i \leq U^a_{g-f} = pf \qquad (i\in I)$$

or

$$(**) \qquad U^a_{f_i} \leq u_i + pf \qquad (i\in I).$$

Define

$$Y := \{y\in\Omega : g(y)-f(y) > 0\} \text{ and } Y_i := \{y\in\Omega : f_i(y) > 0\},\ (i\in I),$$

hence

$$Y_i \subset Y \quad \text{for all} \quad i\in I.$$

If $y\in Y$, then $f(y) < g(y) \leq 1$, hence the inequality

$$U^a_{f_i} \leq u_i + p$$

holds a.e.[m] on Y_i for all $i\in I$ by (**), thus by the complete maximum principle a.e.[m] on Ω, and finally,

$$U^a_{g-f} \leq p \quad \text{a.e.[m] on } \Omega.$$

Hence by (*),

$$pR^a_p g = U_{g-f} \leq p \quad \text{or} \quad R^a_p g \leq 1 \quad \text{a.e.[m] on } \Omega.$$

(4) $\Rightarrow$ (1): We shall prove the more general assertion. Let T be the projection of $\mathbb{R}$ onto $[a,b]$ with $a\leq 0\leq b$ and $u\in\mathcal{H}$. Then $T\circ u\in\mathcal{H}$ and

$$a(u+T\circ u,\ u-T\circ u) \geq 0.$$

First, let $f\in\mathcal{C}_c(\Omega)\cap\mathcal{H}$, then $T\circ f\in\mathcal{C}_c(\Omega)$, and using the second formula of 2) of (5.3), we get the following estimates

$$H^a_p(T\circ f,\ T\circ f)$$

$$= \frac{1}{2p}\Big[\int(2-m^a_p-\hat m^a_p)\,|T\circ f|^2 dm + \iint|T\circ f(x)-T\circ f(y)|^2\sigma^a_p(dx,dy)$$

$$\leq \frac{1}{2p}\Big[\int(2-m^a_p-\hat m^a_p)\,f^2 dm + \iint|f(x)-f(y)|^2\sigma^a_p(dx,dy) = H^a_p(f,f)\ ,$$

i.e. $T \circ f \in \mathcal{C}_c(\Omega) \cap \mathcal{H}$ according to (5.2) and

$$a(T\circ f,\ T\circ f) \leq a(f,f)$$

by (5.1). By the first formula of 2) of (5.3), we get

$$H_p^a(f - T\circ f,\ T\circ f)$$
$$= \frac{1}{p}[\int(1-m_p^a)(f-T\circ f)T\circ f\ dm + \iint[f(x)-T\circ f(x)][T\circ f(x)-T\circ f(y)]\sigma_p^a(dx,dy)$$
$$= \frac{1}{p}[I_1 - I_2].$$

To show: $I_1 \geq 0$ and $I_2 \geq 0$.

Since

$$[f(x)-T\circ f(x)]T\circ f(x) = \begin{cases} (f(x)-a)a, & \text{if} \quad f(x) \leq a \\ 0\quad , & \text{if} \quad a \leq f(x) < b \\ (f(x)-b)b, & \text{if} \quad f(x) \geq b \end{cases}$$

and $a \leq 0 \leq b$, we get $I_1 \geq 0$.

Now let

$$A: = [f \leq a] \times \Omega, \qquad B: = [f \geq b] \times \Omega.$$

Then

$$I_2=\iint_A[f(x)-a][a-T\circ f(y)]\sigma_p^a(dx,dy)+\iint_B[f(x)-b][b-T\circ f(y)]\sigma_p^a(dx,dy) \geq 0,$$

since $a \leq T\circ f(y) \leq b$, hence

$$H_p^a(f-T\circ f,\ T\circ f) \geq 0$$

which implies by (5.1)

$$a(f-T\circ f,\ T\circ f) \geq 0.$$

Using this inequality, we get the desired result:

$$a(f-T\circ f,f+T\circ f) = a(f-T\circ f,f-T\circ f) + 2a(f-T\circ f,T\circ f) \geq 0.$$

In an analogous manner, we get the inequality

$$a(f+T\circ f,\ f-T\circ f) \geq 0.$$

The assertion holds now for an arbitrary $u \in \mathcal{H}$ by approximation in the same way as in the proof ((5) $\Longrightarrow$ (1)) of (6.1). $\lrcorner$

(8.2) <u>COROLLARY</u>: Let a be a Dirichlet form on $\mathcal{H}$, and let T be the projection of $\mathbb{R}$ onto a closed interval containing 0. Then T operates on $\mathcal{H}$ with respect to a and $\hat{a}$. $\lrcorner$

(8,3) COROLLARY: Let a be a Dirichlet form on $\mathcal{H}$. Then $0 \leq f \leq 1$ for $f \in \mathcal{H} \cup L^2$ implies $0 \leq R_p^a f \leq 1$ and $0 \leq \hat{R}_p^a \leq 1$ for all $p > 0$.

9. Representations of Dirichlet forms

In this section, we shall extend a representation theorem of A.BEURLING - J. Deny ([2], p. 212) for a Dirichlet form a on a regular functional space $\mathcal{H} = \mathcal{H}(\Omega,m)$. We denote by α (resp. β) the symmetric (resp. anti-symmetric) part of a.

(9.1) LEMMA: Provided with the norm

$$\|\cdot\|_\alpha := [\alpha(\cdot,\cdot)]^{\frac{1}{2}} \quad ,$$

$\mathcal{H}$ is a Dirichlet space, i.e. a regular functional space such that all normalized contractions operate on $\mathcal{H}$ with respect to the inner product.

Proof: By [7]. p.22, it is sufficient, that the unit contraction operates on $\mathcal{H}$, which is clear by the definition of a Dirichlet form. $\lrcorner$

(9.2) THEOREM: For every $f, g \in \mathcal{C}_c(\Omega) \cap \mathcal{H}$, we have the following representations:

$$\alpha(f,g) = \iint fg\, d\nu_\alpha + \iint [f(x)-f(y)][g(x)-g(y)]\sigma^\alpha(dx,dy) + N_\alpha(f,g) \;;$$

$$\beta(f,g) = \iint f(x)g(y)\sigma^\beta(dx,dy) + N_\beta(f,g) \quad ,$$

where:

(1) $\nu_\alpha = \lim_{p\to o} \frac{1}{2p} [2 - m_p^a - \hat{m}_p^a]m$ (in the vague topology) ;

(1) $\sigma^\alpha = \frac{1}{2} (\hat{\sigma}^a + \sigma^a)$ on $\Omega \times \Omega \setminus \Delta$;

(3) $\sigma^\beta = \frac{1}{2} (\hat{\sigma}^a - \sigma^a)$ on $\Omega \times \Omega \setminus \Delta$;

(4) N_α is a positive Hermitian form, and N_β an anti-symmetric

bilinear form, N_α and N_β have the following local character:

If g is constant in some neighborhood of the support $S(f)$ of f, then $N_\alpha(f,g) = N_\beta(f,g) = 0$.

The above representation is unique.

<u>Proof:</u> Since $\mathcal{H}$ is a Dirichlet space with respect to the norm $\|\cdot\|_\alpha$, we get by [2], p. 212

$$\alpha(f,g) = \iint fg\,d\nu_\alpha + \iint [f(x)-f(y)][g(x)-g(y)]\sigma^\alpha(dx,dy) + N_\alpha(f,g) \quad ,$$

where σ^α is the singular measure of α and ν_α a positive measure on Ω . By [8], p 339, this representation is unique.

Using the first formula of 2) of (5.3), we can write α as

$$\alpha(f,g) = \lim_{p\to o} \tfrac{1}{2} [H_p^a(f,g) + H_p^a(g,f)]$$

$$= \lim_{p\to o} \frac{1}{2p}[\int(2-m_p^a-\hat{m}_p^a)fgdm+\tfrac{1}{2}\iint[f(x)-f(y)][g(x)-g(y)][\hat{\sigma}_p^a+\sigma_p^a](dx,dy)]$$

$$= \lim_{p\to o} \frac{1}{2p}\int(2-m_p^a-\hat{m}_p^a)fgdm+\tfrac{1}{2}\iint[f(x)-f(y)][g(x)-g(y)][\hat{\sigma}^a+\sigma^a](dx,dy)$$

$$+N_\alpha'(f,g),$$

where

$$N_\alpha'(f,g) = \lim_{p\to o} \frac{1}{2p} \iint[f(x)-f(y)][g(x)-g(y)][\hat{\sigma}_p^a + \sigma_p^a](dx,dy)$$

$$- \tfrac{1}{2} \iint[f(x)-f(y)][g(x)-g(y)][\hat{\sigma}^a + \sigma^a](dx,dy).$$

Obviously, $\frac{1}{2p}(\hat{\sigma}_p^a + \sigma_p^a)$ converges vaguely to σ^α on $\Omega \times \Omega \setminus \Delta$ by the definition of the singular measures. Hence we have (2) and therefore $N_\alpha' = N_\alpha$. By the uniqueness of the above representation for α, we get (1). Since $\beta(f,g) = \frac{1}{a}a(f,g) - \frac{1}{2}a(f,g) =$

$$\lim_{p\to o} \frac{[H_p^a(f,g)-H_p^a(g,f)]}{2} = \lim_{p\to o} \frac{1}{2p} \iint f(x)g(y)[\hat{\sigma}_p^a-\sigma_p^a](dx,dy) \quad ,$$

the bilinear form N_β, defined by

$$N_\beta(f,g) = \lim_{p\to o} \frac{1}{2p} \iint f(x)g(y)[\hat{\sigma}_p^a-\sigma_p^a](dx,dy)-\iint f(x)g(y)\sigma^\beta(dx,dy),$$

is well defined where σ^β is given by (3). Evidently, N_β has

property (4), thus $\beta(f,g) = \iint f(x)g(y)\sigma^{\beta}(dx,dy) + N_{\beta}(f,g)$.
The proof of the uniqueness of the representation for β is as follows: Suppose that there exists another measure τ^{β} on $\Omega \times \Omega \setminus \Delta$ and another anti-symmetric bilinear form N'_{β} which satisfies (4) such that

$$\beta(f,g) = \iint f(x)g(y)\tau^{\beta}(dx,dy) + N'_{\beta}(f,g).$$

Let $f \otimes g \in \mathfrak{F}_{\Delta}$. Then $N'_{\beta}(f,g) = N_{\beta}(f,g) = 0$, hence

$$\beta(f,g) = \iint f(x)g(y)\tau^{\beta}(dx,dy) = \iint f(x)g(y)\sigma^{\beta}(dx,dy) ,$$

which implies $\tau^{\beta} = \sigma^{\beta}$ because $\mathfrak{F}_{\Delta}$ is a total class in $\mathcal{C}_c(\Omega \times \Omega\setminus\Delta)$. Therefore $N'_{\beta} = N_{\beta}$ on $\mathcal{C}_c(\Omega) \cap \mathcal{H}$. This completes the proof. ⌟

10. Examples

The following example is based on the work by G.STAMPACCHIA [12]. Let Ω be an open, relatively compact subset of $\mathbb{R}^n (n > 2)$ and $\bar{\Omega}$ be its closure. The restriction to Ω of the n-dimensional Lebesgue measure is denoted by m, and the L^p-norm with respect to m by $\|\cdot\|_p$ $(0 < p)$.
$\mathcal{C}'_o(\Omega)$ is the space of all (real-valued) continuous functions on Ω which have continuous partial derivatives of the first order and a compact support in Ω. The completion of $\mathcal{C}'_o(\Omega)$ normed by

$$\|u\|_o := \sum_{i=1}^{n} \|u_{x_i}\|_2$$

is a Hilbert space, denoted by $\mathcal{H}$.
According to Sobolev's lemma ([12], lemma 1.3), there exists a constant S depending only on n, such that for all $u \in \mathcal{H}$

$$\|u\|_q \leq S \cdot \sum_{i=1}^{n} \|u_{x_i}\|_2 = S \|u\|_o ,$$

where $\frac{1}{q} = \frac{1}{2} - \frac{1}{n}$. Hence $\mathcal{H}$ is a regular functional space with respect to Ω and m.

Consider the differential operator

(10.1) $$Lu = -\sum_{j=1}^{n} \left(\sum_{i=1}^{n} a_{ij} u_{x_i} + d_j u \right)_{x_j} + \left(\sum_{i=1}^{n} b_i u_{x_i} + cu \right)$$

where a_{ij} are (real-valued) bounded, measurable functions on Ω. Further, we suppose the following conditions:

(10.2) L is uniformly elliptic, i.e. there exists a constant $\nu > 0$ such that

$$\nu \cdot \sum_{i=1}^{n} y_i^2 \leq \sum_{i,j=1}^{n} a_{ij}(x) y_i y_j \qquad (x \in \Omega,\ y \in \mathbb{R}^n)$$

(10.3) $$\begin{cases} |a_{ij}| \leq M \quad ; & (i,j = 1,\dots,n) \\ b_i,\ d_i \in L^n(\Omega,m) \quad ; & (i = 1,\dots,n) \\ c \in L^{\frac{n}{2}}(\Omega,n) \quad , & \end{cases}$$

(10.4) $$S \cdot \sum_{i=1}^{n} \|d_i\|_n + S \cdot \sum_{i=1}^{n} \|b_i\|_n + S^2 \cdot \|c\|_{\frac{n}{2}} < \frac{\nu}{2}$$

(10.5) $$c - \sum_{i=1}^{n} (d_i)_{x_i} \leq 0, \quad c - \sum_{i=1}^{n} (b_i)_{x_i} \leq 0$$ (in the sense of distributions).

Now we associate to L a bilinear form on $\mathcal{H} \times \mathcal{H}$ by

(10.6) $$a(u,v) = \int_{\Omega} \{ \sum_{j=1}^{n} (\sum_{i=1}^{n} a_{ij} u_{x_i} + d_j u) v_{x_j} + (\sum_{i=1}^{n} b_i u_{x_i} + cu) v \} dm \ .$$

We remark that the adjoint operator $\hat{L}$ of L, whose associated bilinear form is the adjoint bilinear form $\hat{a}$ of a, is given by

$$\hat{L}u = -\sum_{j=1}^{n} \left(\sum_{i=1}^{n} a_{ji} u_{x_i} + b_j u \right)_{x_j} + \left(\sum_{i=1}^{n} d_i u_{x_i} + cu \right).$$

(10.7) THEOREM: Let L be the differential operator given by (10.1) and a the bilinearform associated to L by (10.6). If the conditions (10.2)-(10.5) are satisfied then a is a Dirichlet form on $\mathcal{H}$.

Proof: The conditions (10.2) and (10.3) imply that a is a continuous bilinear form on $\mathcal{H} \times \mathcal{H}$ ([12], lemma 1.5). Using (10.4), we get

$$a(u,u) \geq \frac{\nu}{2} \|u\|_o^2 \qquad (u \in \mathcal{H})$$

by [12], théorème 3.1. By théorème 3.5 of [12], the cones $\mathcal{P}^a$ and $\hat{\mathcal{P}}^a$ satisfy the principle of the lower envelope. Finally, the condition (10.5) guarantees that $\inf(u,1) \in \mathcal{P}^a$ (resp. $\in \hat{\mathcal{P}}^a$) for every $u \in \mathcal{P}^a$ (resp. $\in \hat{\mathcal{P}}^a$) ([12], corollaire du théorème 3.5). Since

$$\inf(u,v+1) = \inf(u,v+\inf(u,1)) ,$$

$\mathcal{P}^a$ and $\hat{\mathcal{P}}^a$ both satisfy the strong principle of the lower envelope, hence a is a Dirichlet form by (8.1). $\lrcorner$

II. Potential theory of Dirichlet forms

Throughout this chapter, let a be a Dirichlet form on a regular functional space $\mathcal{H} = \mathcal{H}(\Omega,m)$. We shall show, that the cones $\mathcal{P}^a$ and $\hat{\mathcal{P}}^a$ satisfy the main potential-theoretic principles which are known in classical potential theory.

11. The principle of the convex envelope

DEFINITION: An open set $\omega \subset \Omega$ is called a-regular for $u \in \mathcal{H}$ if $a(u,\varphi) = 0$ for all $\varphi \in \mathcal{C}_c(\omega) \cap \mathcal{H}$.

(11.1) LEMMA: Let ω_1 and ω_2 be a-regular for $u \in \mathcal{H}$. Then $\omega_1 \cup \omega_2$ is a-regular for u.

Proof: Let $\varphi \in \mathcal{C}_c(\omega_1 \cup \omega_2) \cap \mathcal{H}$. To show: $a(u,\varphi)= 0$. Define $K = S(\varphi) \cap \complement\omega_1$, and let K' be a compact neighborhood of K such that $K' \subset \omega_2$. There exists a function $f \in \mathcal{C}_c^+(\Omega) \cap \mathcal{H}$ such that $f(x) = 1$ on K', $f(x) = 0$ on $\complement\omega_2$. Indeed, by the regularity of $\mathcal{H}$, we can find a function $f' \in \mathcal{C}_c^+(\Omega) \cap \mathcal{H}$ such that $f(x) \geq 1$ on K' and $f(x) \leq 0$ on $\complement\omega_2$. Then $f = T_I \circ f'$ does the job. Since $\mathcal{H}$, provided with the norm $\|\cdot\|_\alpha$, is a Dirichlet space by (9.1), we have by [2], p. 213, $f\cdot\varphi \in \mathcal{C}_c(\Omega)\cap\mathcal{H}$. By the definitions of f and φ, we get

$$(f\cdot\varphi)(x) = 0 \text{ on } \complement\omega_2,\ \varphi(x) - (f\cdot\varphi)(x) = 0 \text{ on } \complement\omega_1,$$

hence

$$a(u,\varphi) = a(u,f\cdot\varphi) + a(u,\varphi - f\varphi) = 0.$$ _/

<u>DEFINITION</u>: A point $x \in \Omega$ is called <u>a-regular</u> for $u \in \mathcal{H}$ if there exists an open neighborhood of x which is a-regular for u.

(11.2) <u>LEMMA</u>: The set of all $x \in \Omega$ which are a-regular for $u \in \mathcal{H}$ is the greatest a-regular set for u.

<u>Proof</u>: Let $(\omega_i)_{i\in I}$ be the family of all a-regular sets for u. To show: $\omega := \bigcup_{i\in I}\omega_i$ is a-regular for u.
Let $\varphi \in \mathcal{C}_c(\omega)\cap\mathcal{H}$. Since $S(\varphi)$ is compact, there exist $\omega_1,\dots,\omega_n \in (\omega_i)_{i\in I}$ such that $S(\varphi) \subset \bigcup_{i=1}^n \omega_i$.
Since by (11.1) $\bigcup_{i=1}^n \omega_i$ is a-regular for u, we get $a(u,\varphi) = 0$. _/

<u>DEFINITION:</u> The complement of the greatest open a-regular set for $u \in \mathcal{H}$ is called the <u>a-spectrum</u> of u and denoted by $\Sigma^a(u)$. In an analogous manner, the <u>a-cospectrum</u> $\hat{\Sigma}^a(u)$ of u is defined. We define for any open set $\omega \subset \Omega$:

$$W^a_\omega := \overline{\{u \in \mathcal{H} : \Sigma^a(u) \subset \omega\}}.$$

(11.3) <u>REMARKS</u>: 1) $\Sigma^a(u)$ is closed, and $\Sigma^a(u) = \emptyset$ iff $u = 0$.

2) If $U^a_\mu \in \mathcal{P}^a$, then $\Sigma^a(U^a_\mu) = S(\mu)$, where $S(\mu)$ denotes the support of μ.

3) If $\Sigma^a(u) \subset \omega$, then $a(u,\varphi) = 0$ for all $\varphi \in \mathcal{C}_c(\Omega)\cap\mathcal{H}$ such that $S(\varphi)\cap\omega = \emptyset$, hence this property holds for all $u \in W^a_\omega$ by the continuity of a. Therefore, $u \in W^a_\omega$

implies $\Sigma^a(u) \subset \overline{\omega}$. More precisely, we have the following lemma:

(11.4) <u>LEMMA</u>: Let $u \in W^a_\omega$. Then $a(u,v) = 0$ for all $v \in \mathcal{H}$ such that $v = 0$ a.e.[m] on ω.

<u>Proof</u>: It suffices to prove $a(u,v) = 0$ for all $u,v \in \mathcal{H}$ such that $\Sigma^a(u) \subset \omega$ and $v = 0$ on ω. We shall use the following fact: If $(w_n) \subset \mathcal{H}$ converges strongly in $\mathcal{H}$ to $w \in \mathcal{H}^+$, then $(|w_n|)$ converges strongly in $\mathcal{H}$ to $|w| = w$. Indeed, using (9.1), the convergence follows from [4], exposé no. 1, p. 4.

1) Let $v \in \mathcal{C}_c(\Omega) \cap \mathcal{H}$ with $S(v) \subset \complement\Sigma^a(u)$. Then by definition,

$$a(u,v) = 0 .$$

2) Let $v \in \mathcal{H}^+$ be bounded with compact support. There exists $f \in \mathcal{C}^+_c(\Omega) \cap \mathcal{H}$ such that $f \geq v$ and $S(f) \cap \Sigma^a(u) = \emptyset$. Furthermore, let (φ_n) be a sequence in $\mathcal{C}^+_c(\Omega) \cap \mathcal{H}$ which converges strongly in $\mathcal{H}$ to v. If

$$\psi_n = \inf(f,\varphi_n) = \frac{1}{2}(f+\varphi_n) - \frac{1}{2}|f-\varphi_n|,$$

then $\psi_n \in \mathcal{C}^+_c(\Omega) \cap \mathcal{H}$ such that $S(\psi_n) \cap \Sigma^a(u) = \emptyset$.
Since $(f + \varphi_n)$ converges strongly in $\mathcal{H}$ to $f + v$ and $(|f - \varphi_n|)$ converges strongly in $\mathcal{H}$ to $|f-v| = f-v$, we get strong convergence in $\mathcal{H}$ of (ψ_n) to v. By (1), we have $a(u,\psi_n) = 0$, hence $a(u,v) = 0$.

3) Let $v \in \mathcal{H}^+$ such that $S(v) \cap \Sigma^a(u) = \emptyset$, and again, let (φ_n) be a sequence in $\mathcal{C}^+_c(\Omega) \cap \mathcal{H}$ which converges strongly in $\mathcal{H}$ to v. Define $\psi_n = \inf(\varphi_n, v)$. Then $\psi_n \geq 0$, bounded with compact support such that $S(\psi_n) \cap \Sigma^a(u) = \emptyset$. Since (ψ_n) converges weakly to v, we get by (2) $a(u,v) = 0$.

4) Let $v \in \mathcal{H}$ such that $v = 0$ on ω. Writing v as a linear combination of elements in $\mathcal{H}^+$ which vanish on ω, we get $a(u,v) = 0$ by (3). $\lrcorner$

(11.5) LEMMA: Let T be a normalized contraction which is given as a projection of $\mathbb{R}$ onto a closed interval containing 0. If $u \in W^a_\omega$, then $u = T \circ u$ a.e.[m] on ω implies $u = T \circ u$.

Proof: By (8.2) $T \circ u \in \mathcal{H}$ and $a(u+T\circ u, u-T\circ u) \geq 0$. Since $u-T\circ u = 0$ a.e.[m] on ω, we have $a(u,u-T\circ u) = 0$ by (11.4), hence

$$0 \leq c \cdot \|u-T\circ u\|^2 \leq a(u-T\circ u, u-T\circ u)$$
$$= a(u,u-T\circ u) - a(T\circ u, u-T\circ u) \leq 2a(u,u-T\circ u) = 0 ,$$

thus $u = T\circ u$. _/

(11.6) THEOREM (Principle of the convex envelope): Let $u \in W^a_\omega$. Then the image $u(\Omega)$ of Ω under u is contained in the smallest closed interval containing $u(\omega) \cup \{0\}$.

Proof: Let T be the projection of $\mathbb{R}$ onto the smallest closed interval containing $u(\omega) \cup \{0\}$. Then $u = T\circ u$ a.e.[m] on ω, hence by (11.5) $u = T\circ u$.

(11.7) COROLLARY: Let $u \in \mathcal{H}$ be continuous with compact a-spectrum $\Sigma^a(u)$. Then $u(\Omega)$ is contained in the smallest interval containing $\Sigma^a(u) \cup \{0\}$.

(11.8) COROLLARY (Maximum principle): If $u \in W^a_\omega$, then for locally m-almost all $x \in \Omega$

$$|u(x)| \leq \sup_{y\in\omega} |u(y)| .$$

12. Theorem of spectral synthesis

(12.1) LEMMA: The a-projection of a pure a-potential on W^a_ω is a pure a-potential.

Proof: (1) If v' is the $\hat{a}$-projection of an element $v \in \mathcal{H}$ to W^a_ω, then $v = v'$ a.e.[m] on ω. Indeed, v' is characterized by

$$\hat{a}(v', w-v') \geq \hat{a}(v, w-v') \quad \text{for all} \quad w \in W_\omega^a .$$

Since W_ω^a is a closed linear subspace of $\mathcal{H}$, we get for all $w \in W_\omega^a$

$$\hat{a}(v',w) = \hat{a}(v,w) .$$

If $f \in M_c(\omega)$, then $U_f^a \in W_\omega^a$, hence

$$\int v'f \, dm = a(U_f^a, v') = \hat{a}(v', U_f^a)$$
$$= \hat{a}(v, U_f^a) = a(U_f^a, v) = \int fv \, dm,$$

thus $v = v'$ a.e.[m] on ω.

(2) If $v \in \mathcal{H}^+$, then $v' \geq 0$. In fact, the principle of the convex envelope (11.6) yields $v' \geq 0$ since $v' = v \geq 0$ a.e.[m] on ω.

(3) Let $u \in \mathcal{P}^a$, and let u' be the a-projection of u onto W_ω^a. Furthermore, let $v \in \mathcal{H}^+$, and let v' be the $\hat{a}$-projection of v onto W_ω^a.

To show by (2.2): $a(u',v) \geq 0$.

By (11.4) and (1), we get $a(u', v-v') = 0$, hence

$$a(u',v) = a(u',v') .$$

u' is characterized by $a(u',w) = a(u,w)$ for all $w \in W_\omega^a$. Taking $w = v'$, we get $a(u',v') = a(u,v')$, hence

$$a(u',v) = a(u',v') = a(u,v') \geq 0 \quad ,$$

since $u \in \mathcal{P}^a$ and $v' \geq 0$ by (2). $\_/$

(12,2) THEOREM (of spectral synthesis): Every element $u \in \mathcal{H}$ is the strong limit in $\mathcal{H}$ of linear combinations of pure a-potentials whose associated measures are supported by the a-spectrum $\Sigma^a(u)$ of u.

Proof: Define for closed $F \subset \Omega$

$$W_F^a = \{u \in \mathcal{H} : \Sigma^a(u) \subset F\} .$$

Since $F = \bigcap_{\substack{F \subset \omega \\ \omega \text{ open}}} \overline{\omega}$, we get by remark (3) of (11.3)

$$W_F^a = \bigcap_{\substack{F \subset \omega \\ \omega \text{ open}}} W_\omega^a \ .$$

Using (1.2) and (12.1), we see that the a-projection of a pure a-potential onto W_F^a is again a pure a-potential. Since the linear combinations of pure a-potentials are dense in $\mathcal{H}$, the a-projections onto W_F^a of these linear combinations are dense in W_F^a, hence the assertion. _/

13. The condensor principle

(13.1) LEMMA: To each element $u \in \mathcal{H}$ it is possible to associate a function u^*, called refinement of u, such that

(1) $u^* = u$ a.e.[m] on Ω.

(2) u^* is measurable with respect to the measure associated to any pure a-potential U_μ^a; and

$$a(U_\mu^a, a) = \int u^* d\mu \ .$$

Proof: By (9.1) $\mathcal{H}$, provided with the norm $\|\cdot\|_\alpha$, is a Dirichlet space, and the assertion follows from [2], p. 210._/

(13,2) THEOREM (Condensor principle): Let ω_0 and ω_1 be two open sets of Ω with disjoint closures, ω_1 to be relatively compact. Then there exists an a-potential $U_{\mu-\nu}^a$ where μ and ν are positive measures on Ω such that

(1) $0 \leq U_{\mu-\nu}^a \leq 1$ a.e.[m] on Ω;

(2) $U_{\mu-\nu}^a = 0$ a.e.[m] on ω_0, $U_{\mu-\nu}^a = 1$ a.e.[m] on ω_1;

(3) μ is supported by $\overline{\omega}_1$, ν by $\overline{\omega}_0$;

(4) $\int d(\mu-\nu) \geq 0$.

Proof: Define a non-empty, closed, convex set $A \subset \mathcal{H}$

by

$$A = \{v \in \mathcal{H} : v \geq 1 \text{ a.e.}[m] \text{ on } \omega_1,\ v \leq 0 \text{ a.e.}[m] \text{ on } \omega_o\} ,$$

and let u be the a-projection of 0 onto A: $u = \pi_A^a(0)$; u is characterized by

$$a(u,v-u) \geq 0 \quad \text{for all} \quad v \in A .$$

Define $u' = T_I \circ u$. Since $u' \in A$, we have $a(u,u'-u) \geq 0$. On the other hand, T_I operates on $\mathcal{H}$ with respect to a. This implies $a(u+u', u-u') \geq 0$ or $a(u',u-u') \geq a(u,u'-u)$, hence

$$0 \leq c\|u-u'\|^2 \leq a(u-u', u-u') = a(u,u-u') - a(u',u-u') \leq 0 ,$$

i.e. $u = u' = T_I \circ u$. Therefore u satisfies the conditions (1) and (2). Next we shall show, that u is an a-potential. Define:

$$A_o := \{v \in \mathcal{C}_c(\Omega) \cap \mathcal{H} : v \geq 0 \text{ on } \omega_1,\ S(v) \subset \complement\overline{\omega}_o\} ,$$

$$A_1 := \{w \in \mathcal{C}_c(\Omega) \cap \mathcal{H} : w \geq 0 \text{ on } \omega_o,\ S(w) \subset \complement\overline{\omega}_1\} .$$

If $t > 0$ and $v \in A_o$, then $u + tv \in A$, hence

$$0 \leq a(u,(u+tv)-u) = ta(u,v) \quad \text{or} \quad a(u,v) \geq 0 .$$

Thus there exists a positive measure μ, sopported by $\overline{\omega}_1$, such that

$$a(u,v) = \int f \, d\mu \quad \text{for all} \quad v \in A_o .$$

In an anlogous manner, a positive measure ν, supported by $\overline{\omega}_o$, exists such that

$$a(u,w) = -\int w \, d\nu \quad \text{for all} \quad w \in A_1 .$$

Now let $\varphi \in \mathcal{C}_c^+(\Omega) \cap \mathcal{H}$, and choose $v', v'' \in A_o$, $w', w'' \in A_1$ such that

$$v' \leq \varphi \leq v'' \text{ on } \omega_1 \quad \text{and} \quad w' \leq \varphi \leq w'' \text{ on } \omega_o .$$

Since for all $t > 0$

$$u+t(v''+w'-\varphi) \in A \quad \text{and} \quad u+t(\varphi-v'-w'') \in A ,$$

we get $a(u,v''+w'-\varphi) \geq 0$ and $a(u,\varphi-v'-w'') \geq 0$, which implies $a(u,v'+w'') \leq a(u,\varphi) \leq a(u,v''+w;)$,

hence $\int v'd\mu - \int w''d\nu \leq a(u,\varphi) \leq \int v''d\mu + \int w'd\nu$.

If v' and v'' converge uniformly on ω_1 to φ, and w' and w'' converge uniformly on ω_o to φ, the last estimates imply

$$a(u,\varphi) = \int \varphi d\mu - \int \varphi d\nu = \int \varphi d(\mu-\nu) ,$$

hence by (6.3)

$$u = U^a_{\mu-\nu} .$$

Still to show (4). Let ω be open and relatively compact such that $\overline{\omega}_1 \subset \omega$. If we apply this theorem to the sets ω and $\emptyset$, then there exists a pure a-potential U^a_τ such that

$0 \leq U^a_\tau \leq 1$ a.e.[m] on Ω, $U^a_\tau = 1$ a.e.[m] on ω, and the positive measure τ is supported by $\overline{\omega}$. Since ω is relatively compact, $(U^a_\tau)^*$ is $(\mu-\nu)$-measurable by [7], p.13, Using property (1) and the characteristic property (2,2) of a pure a-potential, we get

$$0 \leq a(U^a_{\mu-\nu},U^a_\tau) = \int (U^a_\tau)^* d(\mu-\nu) = \int (U^a_\tau)^* d\mu - \int (U^a_\tau)^* d\nu$$
$$\leq \int d\mu - \int_\omega (U^a_\tau)^* d\nu = \int d\mu - \int d\nu .$$

ω being arbitrary, we get $\int d(\mu-\nu) \geq 0$. ⌟

DEFINITION: The a-potential $U^a_{\mu-\nu}$ of the proof of (13.2) is called the a-condensor potential of ω_1 and ω_o and $\mu-\nu$ is the a-condensor measure of ω_1 and ω_o.

Taking $\omega_o = \emptyset$ in (13,2), we get the following corollary:

(13.3) COROLLARY (Equilibrium principle): For any open relatively compact set $\omega \subset \Omega$, there exists a pure a-potential U^a_μ such that

(1) $0 \leq U^a_\mu \leq 1$ a.e.[m] on Ω;

(2) $U^a_\mu = 1$ a.e.[m] on ω;

(3) μ is supported by $\overline{\omega}$.

DEFINITION: The pure a-potential U_μ^a of (13.3) is called the a-equilibrium potential of ω and μ the a-equilibrium measure of ω.

(13.4) REMARK: By the above definition and by [3], the a-equilibrium potential and the a-capacitary potential of ω coincide. Furthermore,

$$\mathrm{cap}_a \omega = \int d\mu = \mu(\overline{\omega}) ,$$

where the a-capacity of ω is defined by $\mathrm{cap}_a \omega = a(U_\mu^a, U_\mu^a)$. Indeed, let $\varphi \in \mathcal{C}_c^+(\Omega) \cap \mathcal{H}$ such that $\varphi = 1$ on $\overline{\omega}$. Then

$$\mu(\overline{\omega}) = \int d\mu = \int \varphi d\mu = a(U_\mu^a, \varphi) \geq a(U_\mu^a, U_\mu^a) = \mathrm{cap}_a \omega,$$

where the inequality follows from the definition of U_μ^a . On the other hand, there exists a net $(f_i) \subset M_c^+(\omega)$ such that $(U_{f_i}^a)$ converges strongly in $\mathcal{H}$ to U_μ^a, hence the net of measures $(f_i m)$ converges vaguely to μ, hence

$$a(U_{f_i}^a, U_\mu^a) = \int U_\mu^a \cdot f_i dm \geq \int f_i dm .$$

Since $(f_i \cdot m)$ converges vaguely to μ and ω is relatively compact, we obtain

$$\mathrm{cap}_a \omega = a(U_\mu^a, U_\mu^a) \geq \int d\mu .$$ _/

14. Balayage theory

(14.1) THEOREM (Balayage principle): Given a pure a-potential U_μ^a and an open subset ω of Ω, there exists a pure a-potential $U_{\mu'}^a$, such that

(1) $U_{\mu'}^a \leq U_\mu^a$ a.e.[m] on Ω ;

(2) $U_{\mu'}^a = U_\mu^a$ a.e.[m] on ω ;

(3) μ' is supported by $\overline{\omega}$;

(4) $\int d\mu' \leq \int d\mu$.

Proof: Define $u = \pi_A^a(0)$, where A is the following non-empty, closed convex subset of $\mathcal{H}$:

$$A = \{v \in \mathcal{H} : v \geq U_\mu^a \text{ a.e.}[m] \text{ on } \omega\}.$$

The element u is characterized by

$$a(u, v-u) \geq 0 \quad \text{for all} \quad v \in A .$$

If $w \in \mathcal{H}^+$, then $u + w \in A$, hence

$$a(u,w) = a(u,(u+w)-u) \geq 0 ,$$

i.e. $u \in \mathcal{P}^a$. Moreover, there exists a positive measure μ' on Ω such that $u = U_{\mu'}^a$. Since $g = \inf(U_\mu^a, u) \in \mathcal{P}^a \cap A$, we get

$$0 \leq c \cdot \|u-g\|^2 \leq a(u-g,u-g) = a(u,u-g)-a(g,u-g) \leq 0 ,$$

hence $u = g = \inf(U_\mu^a,u) \leq U_\mu^a$. Therefore, $u = U_{\mu'}^a$ satisfies (1) and (2). Let $\varphi \in \mathcal{C}_c(\Omega) \cap \mathcal{H}$ such that $S(\varphi) \cap \overline{\omega} = \emptyset$. Then

$$U_{\mu'}^a + t\varphi \in A \quad \text{for all real} \quad t ,$$

hence $0 \leq a(U_{\mu'}^a, (U_{\mu'}^a + t\varphi) - U_{\mu'}^a) = ta(U_{\mu'}^a, \varphi) = t\int \varphi d\mu'$, which implies $\int \varphi d\mu' = a(U_{\mu'}^a, \varphi) = 0$, i.e. (3).

To prove (4) we take an open relatively compact set ω'. Using the a-equilibrium potential $\hat{U}_\tau^a$ of ω', we get the following estimates

$$\begin{aligned}\int_{\omega'} d\mu' &\leq \int (\hat{U}_\tau^a)^* d\mu' = a(U_{\mu'}^a, \hat{U}_\tau^a) = \int (U_{\mu'}^a)^* d\tau \\ &\leq \int (U_\mu^a)^* d\tau = a(U_\mu^a, \hat{U}_\tau^a) = \int (\hat{U}_\tau^a)^* d\mu \leq \int d\mu .\end{aligned}$$

Since ω' was arbitrary, (4) follows. $\quad\_/$

DEFINITION: The pure a-potential $U_{\mu'}^a$ of the proof of (14.1) is called the a-balayaged potential of U_μ^a with respect to ω, and μ' the a-balayaged measure of μ with respect to ω.

(14.2) THEOREM: Let $\omega \subset \Omega$ be open, and $U_\mu^a \in \mathcal{P}^a$. Then the a-balayaged potential $U_{\mu'}^a$ of U_μ^a with respect to ω is the â-projection of U_μ^a onto $W_\omega^a = \{v \in \mathcal{H} : \Sigma^a(v) \subset \omega\}$.

Proof: Denote the $\hat{a}$-projection of an element $u \in \mathcal{H}$ onto W^a_ω by u'. We have by the proof of (12.1)

$$\hat{a}(u',w) = \hat{a}(u,w) \quad \text{for all} \quad w \in W^a_\omega$$

and $u' = u$ a.e.[m] on ω.

Now let $u = U^a_\mu$ and $v = U^a_{\mu'}$. To show: $u' = v$.

Since $u' = u$ a.e.[m] on ω, we have $u' \in A = \{w \in \mathcal{H} : w \geq u$ a.e.[m] on $\omega\}$.

Consequently, by the definition of v

$$(1) \quad a(v,u'-v) \geq 0.$$

On the other hand, since $u = v$ a.e.[m] on ω, we have by (11.4):

$$(2) \quad a(u',u-v) = 0, \quad \text{i.e.} \quad a(u',u) = a(u',v) .$$

Since $u' = u$ a,e,[m] on ω, we get by (11.4) again

$$(3) \quad a(u',u-u') = 0, \quad \text{i.e.} \quad a(u',u) = a(u',u') ,$$

hence

$$0 \leq c\|u'-v\|^2 \leq a(u'-v,u'-v) = a(u',u'-v)-a(v,u'-v) \underset{(1)}{\leq} a(u',u')-a(u',v) \underset{(2),(3)}{=} 0 ,$$

i.e $u' = v$. $\square$

(14.3) COROLLARY (Transitivity of the balayage): Let ω' and ω'' be open subsets of Ω such that $\omega' \subset \omega''$. For a pure a-potential u denote by u' (resp. u'') the a-balayaged potential of u with respect to ω' (resp. ω''). Then u' is the a-balayaged potential of u'' with respect to ω'.

Proof: Denote by v the a-balayaged potential of u'' with respect to ω'. By (14.2), u'', u', and v are characterized by

$$(1) \quad \hat{a}(u'', w-u'') \geq \hat{a}(u,w-u'') \quad \text{for all} \quad w \in W^a_{\omega''} ;$$

$$(2) \quad \hat{a}(u',w-u) \geq \hat{a}(u, w-u') \quad \text{for all} \quad w \in W^a_{\omega'} ;$$

$$(3) \quad \hat{a}(v, w-v) \geq \hat{a}(u'', w-v) \quad \text{for all} \quad w \in W^a_{\omega'} .$$

Take in (2) $w = v$, in (3) $w = u'$. Then

$$0 \leq c \cdot \|u'-v\|^2 \leq \hat{a}(u'-v,u'-v) = \hat{a}(u',u'-v)-\hat{a}(v,u'-v)$$
$$\leq \hat{a}(u,u'-v)-\hat{a}(u'',u'-v) = \hat{a}(u-u'',u'-v)$$
$$= a(u-u'',(u'-v+u'')-u'') \leq 0$$

by (1) since $u' - v + u'' \in W^a_{\omega''}$ $(W^a_{\omega'} \subset W^a_{\omega''})$.

(14.4) COROLLARY (Left-continuity of the balayage): Let $u \in \mathcal{P}^a$ and $\omega' \subset \Omega$ be open. For every $\varepsilon > 0$, there exists a closed set $F \subset \omega'$ such that the following condition holds:

For every open subset ω'' with $F \subset \omega'' \subset \omega'$, the a-balayaged potentials u'' and u' of u with respect to ω'' and ω' satisfy

$$\|u'' - u'\| < \varepsilon .$$

Proof: Let $\{\omega_i\}_{i \in I}$ be the family of open sets with $\omega_i \subset \omega'$. Since $\underset{i \in I}{\cup} \omega_i = \omega'$ and the family $(W^a_{\omega_i})_{i \in I}$ is upward filtering with $\overline{\underset{i \in I}{\cup} W^a_{\omega_i}} = W^a_{\omega'}$, it follows by (1.2) that the family $(u_i)_{i \in I}$ of a-balayaged potentials of u with respect to ω_i converges strongly in $\mathcal{X}$ to u'.

An important information concerning the support of the a-balayaged measure, is given in

(14.5) THEOREM: The following conditions are equivalent:

(1) For any pure a-potential U^a_μ and any closed neighborhood F of the support $S(\mu)$ of u, let $U^a_{\mu'}$ be the a-balayaged potential of U^a_μ with respect to $\omega = \complement F$. Then

$$S(\mu') \subset \omega^* \quad (= \text{topological boundary of } \omega),$$

and the cone $\hat{\mathcal{P}}^a$ has the same property.

(2) $\sigma^a = \hat{\sigma}^a = 0$.

(3) If for $u \in \mathcal{X}$ and $x_o \in \Omega$ there exists an open neighborhood ω of x_o such that $u = 0$ a.e.[m] on ω, then

$$x_o \notin \Sigma^a(u) \cup \hat{\Sigma}^a(\omega) .$$

Proof: (1) $\Rightarrow$ (3): We take another neighborhood ω' of x_0 such that $\omega' \subset \overline{\omega'} \subset \omega$.

For a pure a-potential U^a_μ, let u' be the $\hat{a}$-projection of U^a_μ to $W^a_{\complement\overline{\omega'}}$. If μ is supported by ω', $\Sigma^a(u')$ is contained in ω by (14.2). Hence

$$\hat{a}(u,\ U^a_\mu - u') = 0\ ,$$

because there exists a sequence $(\varphi_n) \subset \mathcal{C}_c(\Omega) \cap \mathcal{H}$ which converges strongly in $\mathcal{H}$ to u and which satisfies $S(\varphi_n) \subset \complement(\Sigma^a(u') \cup \omega')$. Since every element $\varphi \in \mathcal{C}_c(\Omega) \cap \mathcal{H}$ with $S(\varphi) \subset \omega'$ can be approximated in $\mathcal{H}$ by linear combinations of elements of the form $U^a_\mu - u'$, we get $a(u,\varphi) = 0$, hence $x_0 \notin \Sigma^a(u)$.

(3) $\Rightarrow$ (2): Suppose that $\sigma^a \neq 0$. Then there exists $f \otimes g \in \mathcal{F}_\Delta$ such that

$$\iint f(x)g(y)\sigma^a(dx,dy) > 0\ .$$

Since $S(g) \subset \complement S(f)$, we have by (3) $S(g) \subset \complement\Sigma^a(f)$, hence

$$\iint f(x)g(y)\sigma^a(dx,dy) = -\frac{1}{2}\, a(f,g) = 0\ ,$$

a contradiction.

(3) $\Rightarrow$ (1): It suffices to prove

$$\int f\, d\mu' = 0$$

for every $f \in \mathcal{C}^+_c(\Omega) \cap \mathcal{H}$ with $S(f) \subset \omega = \complement F$. Since $f = 0$ in some neighborhood of F, we have $\hat{\Sigma}^a(f) \subset \omega$ by (3).

Without loss of generality, assume that $S(\mu)$ and F are compact. The general case follows by approximation.

We take an open set ω' such that

$$(*) \qquad F \subset \omega' \subset \overline{\omega'} \subset \complement\, \hat{\Sigma}^a(f)\ .$$

By (2) of (14.1), there exists a sequence $(\varphi_n) \subset \mathcal{C}_c(\omega') \cap \mathcal{H}$ which converges strongly in $\mathcal{H}$ to $U^a_\mu - U^a_{\mu'}$. By (*), we have therefore $a(\varphi_n, f) = 0$, hence $0 = a(U^a_\mu - U^a_{\mu'}, f) = \int f d\mu - \int f d\mu' = -\int f d\mu'$.

(2) $\Longrightarrow$ (3): We take another open set ω' such that

$$x_o \in \omega' \subset \overline{\omega'} \subset \omega .$$

There exists a sequence $(\varphi_n) \subset \mathcal{C}_c(\Omega) \cap \mathcal{H}$ with $S(\varphi_n) \subset \complement\,\overline{\omega'}$ which converges strongly in $\mathcal{H}$ to u. Let $f \in \mathcal{C}_c(\omega') \cap \mathcal{H}$. Since $S(\varphi_n) \cap S(f) = \emptyset$, we have by (2):

$$0 = \iint \varphi_n(x) f(y) \sigma^a(dx,dy) = -\tfrac{1}{2}\, a(\varphi_n, f) ,$$

hence, $a(u,f) = 0$ and $x_o \notin \Sigma^a(u)$.

(14.6) EXAMPLE: Let L and a be as in (10.7). If $f,g \in \mathcal{C}_o'(\Omega)$ with $S(f) \cap S(g) = \emptyset$, we get $a(f,g) = 0$. Since these elements $f \otimes g$ form a total class in $\mathcal{C}_c(\Omega \times \Omega \setminus \Delta)$, the singular measures of a and $\hat{a}$ vanish. Hence the condition (1) of (14.5) is given.

15. Bibliography

N.ARONSZAIN-
K.T.SMITH :
[1] Characterization of positive reproducing kernels. Amer. J. Math. 79 (1957), 611-622.

A.BEURLING-
J.DENY : Dirichlet spaces. Proc.Nat.Acad.Sc.45(1959), 208-215.
[2]

J.BLIEDTNER: Functional spaces and their exceptional sets.
[3] Seminar über Potentialtheorie II, Lecture Notes in Math. no.226. Springer-Verlag (1971).

J.DENY : Synthese spectrale dans les espaces de Dirichlet.
[4] Séminaire d'Orsay, 1961/62.

[5] : Principe complet du maximum et contractions. Ann.Inst.Fourier 15 (1965). 259-272.

M.ITO : Characterizations of supports of balayaged measures.
[6] Nagoya Math.J. 28 (1966), 203-230.

[7] : Condensor principle and the unit contraction. Nagoya Math. J. 30 (1967), 9-28.

[8] : The singular measure of a Dirichlet space. Nagoya Math. J. 32 (1968), 337-359.

[9] : A note on extended regular functional spaces. Proc. Jap. Acad. 43 (1967). 435-440.

U.MOSCO : Approximation of the solution of some variational
[10] inequalities. Ann.Sc.Norm.Sup.Pisa 21(1967), 337-394.

G.STAMPACCHIA
[11] : Formes bilinéaires coercitives sur les ensemble convexes. C.R. Acad.Sc. Paris 258 (1064), 4413-4416.

[12] : Le problème de Dirichlet pour les équations elliptiques du second ordre à coefficients discontinus. Ann.Inst.Fourier 15 (1965), 189-259.

K.YOSIDA : Functional analysis. Springer-Verlag New York,(1968).
[13] 2nd edition.

Contents

COHOMOLOGY IN HARMONIC SPACES

by

Wolfhard Hansen

0. Introduction

The investigation of cohomology in harmonic spaces is motivated by the fact that it more or less amounts to studying the index of certain second-order elliptic or parabolic differential operators. To see this consider a differential operator L on a compact differentiable manifold X having the following two properties: 1. L maps the space $\mathcal{E}_X$ of C^∞ functions on X into itself. 2. For every $\varphi \in \mathcal{E}_X$ the equation $Lu = \varphi$ is locally solvable. Denoting by $\mathcal{E}$ the sheaf of germs of local C^∞ functions on X, property (1) implies that L induces a sheaf homomorphism $\mathcal{L}$ of $\mathcal{E}$ into itself which is onto by property (2). Thus we have the exact sequence

$$0 \longrightarrow \ker \mathcal{L} \longrightarrow \mathcal{E} \xrightarrow{\mathcal{L}} \mathcal{E} \longrightarrow 0$$

which is a fine resolution of $\mathcal{H} := \ker \mathcal{L}$ since $\mathcal{E}$ is a fine sheaf.

Hence the cohomology groups of $\mathcal{H}$ are the cohomology groups of the cochain complex

$$0 \longrightarrow \Gamma(X,\mathcal{E}) \xrightarrow{\mathcal{L}'} \Gamma(X,\mathcal{E}) \longrightarrow 0 \longrightarrow \dots$$

where $\Gamma(X,\mathcal{E})$ is the set of all global sections in $\mathcal{E}$ (and canocically isomorphic to $\mathcal{E}_X$), i.e.

$$H^0(X,\mathcal{H}) \cong \ker \mathcal{L}' \cong \ker L,$$
$$H^1(X,\mathcal{H}) \cong \operatorname{coker} \mathcal{L}' \cong \operatorname{coker} L = \mathcal{E}_X / L(\mathcal{E}_X)$$

and $H^q(X,\mathcal{H}) = 0$ for all $q \geq 2$. Therefore the operator L on $\mathcal{E}_X$ is a Fredholm operator if and only if $\dim H^0(X,\mathcal{H}) < \infty$ and

$\dim H^1(X,\mathcal{H}) < \infty$, and its index is the Euler characteristic $\sum_{i=0}^{\infty} (-1)^i \dim H^i(X,\mathcal{H})$ of the sheaf $\mathcal{H}$ of germs of null-solutions of L.

Now suppose that L is a differential operator of the form

$$Lu = \sum_{i=1}^{r} X_i^2 u + Yu + au$$

where a is a C^∞ function on X and $X_1, \ldots, X_r, Y$ are C^∞ vector fields on X such that at every point of X at least one X_i is non-vanishing and the Lie algebra $\mathcal{L}(X_1, \ldots, X_r, Y)$ has rank n (the dimension of the underlying Euclidean space). Then on one hand L has the properties (1) and (2) (Bony, [2]). On the other hand Bony [2] showed that the sheaf $\mathcal{H}$ of germs of null-solutions of L satisfies the axioms of a harmonic space.

But R.M. and M.Hervé ([7], [8]) proved that large classes of second-order elliptic differential operators with discontinuous coefficients lead to harmonic spaces as well.

So a knowledge of cohomology in harmonic spaces may help when dealing with differential operators for which standard methods are difficult or impossible to apply.

B.Walsh ([11],[13]) proved that $H^q(X,\mathcal{H}) = 0$ for every harmonic space $(X,\mathcal{H})$ and all $q \geq 2$, and $\dim H^o(X,\mathcal{H}) = \dim H^1(X,\mathcal{H}) < \infty$ if X is compact. We will give a thorough presentation of this result which in many respects is different from the proof given in [11] and [13].

First we will use the terminology of harmonic kernels introduced in [5]. In fact, the axiomatic theory developed by B.Walsh [13] is merely a special case of the general theory of harmonic kernels in [5]. - It may be noted that the perturbation theory which is interesting in itself and crucial for the second part of the result carries through in the general theory.

Proposition 3.3 (yielding that $\mathcal{R}$ is a fine sheaf) seems to be the only point where it is essential that the harmonic kernels be boundary kernels. - Secondly many proofs and some results are different from those given in [11] or [13]. In particular, we do not need any topology on $\Gamma(X,\mathcal{R})$ or $\Gamma(X,\mathcal{Q})$, but only the supremum norm on the space $\mathcal{B}_X$ of bounded Borel measurable functions on X.

Whereas the basic notions of a harmonic space are introduced in detail, frequent use of the results in [5] is made. The notations of sheaf theory are adopted from [9]. In particular, if $(\mathcal{G}_U, r_U^V)$ is a presheaf of abelian groups over X, $\mathcal{G}$ the associated sheaf (espace étalé) and U open in X, then r_U denotes the canocical mapping of $\mathcal{G}_U$ into the abelian group $\Gamma(U,\mathcal{G})$ of all sections of $\mathcal{G}$ over U , i.e. for each $f \in \mathcal{G}_U$ and $x \in U$ $r_U f(x)$ is the germ $r_U^x f$ of f at x.

1. Local harmonic kernels

Let X be a locally compact space with countable base and $\mathcal{B}$ its σ-algebra of Borel sets. For $A \in \mathcal{B}$ let $\tilde{\mathcal{B}}_A$ denote the linear space of all Borel measurable real-valued functions on A and $\mathcal{B}_A$ the subspace of all bounded functions in $\tilde{\mathcal{B}}_A$. If $A', A \in \mathcal{B}$ with $A' \subset A$ and $f \in \tilde{\mathcal{B}}_A$, the restriction of f to A' will be denoted by $r_A^{A'} f$.

Extending by means of the value 0 we will identify functions defined on any subset of X with functions defined on all of X. For A', $A \in \mathcal{B}$ with $A' \subset A$ we then have $\tilde{\mathcal{B}}_{A'} \subset \tilde{\mathcal{B}}_A$ and for $f \in \tilde{\mathcal{B}}_A$ we get

$$r_A^{A'} f = f \cdot 1_{A'}$$

where $1_{A'}$ denotes the real-valued function having value 1 on A' and value 0 elsewhere.

For our purpose it is the most convenient to consider a finite kernel K on X being a linear mapping from $\mathcal{B}_X$ into $\tilde{\mathcal{B}}_X$ which is positive and σ-continuous (i.e. for which $(f_n) \subset \mathcal{B}_X$ and $f_n \downarrow 0$ implies $Kf_n \downarrow 0$). It is clear that any positive linear mapping from $\mathcal{B}_X$ into $\tilde{\mathcal{B}}_X$ which is dominated by a finite kernel on X is σ-continuous and hence a kernel itself. Let I denote the unit kernel on X, i.e. $If = f$ for all $f \in \mathcal{B}_X$. A kernel K on X is said to be strictly positive, if K1 is strictly positive.

A boundary kernel for an open subset U of X is a finite kernel H_U on X such that the values of $H_U f$ on U depend only on the values of f on the boundary U^* of U, i.e. $H_U 1_{\complement U^*}(x) = 0$ for all $x \in U$, and such that $H_U f(x) = f(x)$ for all $x \in \complement U$ and $f \in \mathcal{B}_X$.

Let $\mathfrak{U}$ be a base of the space X consisting of relatively compact open sets and $\{H_U\}_{U \in \mathfrak{U}}$ a corresponding family of local harmonic kernels, i.e. strictly positive boundary kernels H_U for U having the following properties:

I. For all $f \in \mathcal{B}_X$ the function $H_U f$ is continuous on U and continuous on X, if f is continuous on X.

II. If $V \in \mathfrak{U}$ and $\overline{V} \subset U$, then $H_V H_U = H_U$.

III. ${}^*\mathcal{H}_U^+$ jointly separates the points of U.

If U is an arbitrary open subset of X, then ${}^*\mathcal{H}_U$ denotes the set of all hyperharmonic functions on U, i.e. all functions $s: U \longrightarrow]-\infty,+\infty]$ which are lower semicontinuous and satisfy $H_V s \leq s$ for all $V \in \mathfrak{U}$ with $\overline{V} \subset U$. $\mathcal{H}_U$ is the linear space of all harmonic functions on U, i.e. all continuous real-valued functions h on U satisfying $H_V h = h$ for all $V \in \mathfrak{U}$ with $\overline{V} \subset U$. Obviously $\mathcal{H}_U = {}^*\mathcal{H}_U \cap (-{}^*\mathcal{H}_U)$. $\{H_U\}_{U \in \mathfrak{U}}$ is strong-harmonic, if the locally bounded functions in ${}^*\mathcal{H}_X^+$ jointly separate the points of X.

On every open subset U of X we have an induced family

of local harmonic kernels, namely $\{H_V|\mathcal{B}_U\}_{V\in\mathfrak{U},\ \overline{V}\subset U}$. We say that U is strong-harmonic, if this induced family is strong-harmonic. Each $U \in \mathfrak{U}$ is strong-harmonic by (III).

<u>REMARKS</u>: 1) $\{H_U\}_{U\in\mathfrak{U}}$ is a family of harmonic kernels in the sense of [5]: Consider $x \in X$ and $U \in \mathfrak{U}$ with $x \in U$. Then $h := H_U 1$ is strictly positive, continuous and $H_V h(x) = h(x)$ for every $V \in \mathfrak{U}$ with $\overline{V} \subset U$. Since every H_V is a boundary kernel for V, this implies axiom III of [5].

2) If a subfamily $\mathfrak{U}'$ of $\mathfrak{U}$ is still a base of X, then $\{H_U\}_{U\in\mathfrak{U}'}$ obviously is a family of local harmonic kernels on X which by [5], 5.5 defines the same hyperharmonic and harmonic functions as $\{H_U\}_{U\in\mathfrak{U}}$.

3) It follows from (I), (II) and [5], 5.3 that each $U \in \mathfrak{U}$ is regular with harmonic measures $H_U(x,\cdot)$. Using [5], 5.11, 4.2 we obtain:

a) $\{\mathcal{H}_U, r_U^V|\mathcal{H}_U\}$ is a complete presheaf of linear spaces of continuous real-valued functions.

b) If U is open in X and (h_n) is an increasing sequence in $\mathcal{H}_U$, then $h := \sup h_n \in \mathcal{H}_U$ provided that h is bounded.

c) The regular sets form a base of X.

d) Each point $x \in X$ possesses an open neighborhood U such that there exists a strictly positive $h \in \mathcal{H}_U$ and such that the non-negative hyperharmonic functions on U (with respect to $\{\mathcal{H}_V\}$) jointly separate the points of U.

Conversely, every family $\{\mathcal{H}_U\}$ having the properties (a) - (d) arises from a family of local harmonic kernels. (In (d) it is sufficient to have joint separation by $\mathfrak{W}$-hyperharmonic functions for some family $\mathfrak{W} = \{\mathfrak{W}(x): x \in U\}$ of fundamental systems $\mathfrak{W}(x)$ of regular neighborhoods of x.)

2. Resolution of the sheaf $\mathcal{H}$

For every open subset U of X we choose the following notations: Let $\mathcal{S}_U$ be the convex cone of all continuous real-valued superharmonic functions on U, i.e. all continuous real-valued functions in ${}^*\mathcal{H}_U$ and $\mathcal{P}_U$ be the convex cone of all continuous real-valued potentials on U, i.e. all functions in $\mathcal{S}_U^+$ having greatest harmonic minorant 0. Define

$$Q_U = \mathcal{P}_U - \mathcal{P}_U$$

and

$$\mathcal{R}_U = \mathcal{H}_U + Q_U.$$

Then $\mathcal{R}_U$ and Q_U are $\mathbb{R}$-modules. $\mathcal{R}_U$ is the direct sum of $\mathcal{H}_U$ and Q_U: If $h \in \mathcal{H}_U$ and $p_1, p_2 \in \mathcal{P}_U$ such that $h = p_1 - p_2$, then $h \leq p_1$, hence $h \leq 0$ and $-h \leq p_2$, hence $-h \leq 0$, i.e. $h = 0$.

Denoting by i_U the canonical injection of $\mathcal{H}_U$ into $\mathcal{R}_U$, by π_U the canonical projection of $\mathcal{R}_U$ on Q_U and by j_U the canonical injection of Q_U into $\mathcal{R}_U$, we hence have the exact sequence

$$0 \longrightarrow \mathcal{H}_U \xrightarrow{i_U} \mathcal{R}_U \xrightarrow{\pi_U} Q_U \longrightarrow 0$$

and $\pi_U j_U = \mathrm{id}_{Q_U}$.

It is clear that $\{\mathcal{H}_U, r_U^V|\mathcal{H}_U\}$ is a presheaf of $\mathbb{R}$-modules. Let $\mathcal{H}$ be the corresponding sheaf (espace étalé) called the sheaf of germs of harmonic functions. By [5], 3.1, 5.11 and [9], 4.7, 4.8 the canonical mapping $r_U|\mathcal{H}_U : \mathcal{H}_U \longrightarrow \Gamma(U,\mathcal{H})$ is an isomorphism for every open subset U of X.

For all open subsets V, U of X such that $V \subset U$ we have $r_U^V(\mathcal{H}_U) \subset \mathcal{H}_V$ and $r_U^V(\mathcal{P}_U) \subset r_U^V(\mathcal{S}_U^+) \subset \mathcal{S}_V^+ = \mathcal{H}_V^+ + \mathcal{P}_V \subset \mathcal{R}_V$, hence $r_U^V(\mathcal{R}_U) \subset \mathcal{R}_V$. Thus $\{\mathcal{R}_U, r_U^V|\mathcal{R}_U\}$ is a presheaf of $\mathbb{R}$-modules as well. We denote the corresponding sheaf by $\mathcal{R}$. The canocical mapping of $\mathcal{R}_U$ into $\Gamma(U,\mathcal{R})$ is injective, but not necessarily surjective.

For all open subsets V, U of X such that $V \subset U$ we have the commutative diagram

$$\begin{array}{ccccccccc} 0 & \longrightarrow & \mathcal{H}_U & \xrightarrow{i_U} & \mathcal{R}_U & \xrightarrow{\pi_U} & Q_U & \longrightarrow & 0 \\ & & \downarrow r_U^V & & \downarrow r_U^V & & & & \\ 0 & \longrightarrow & \mathcal{H}_V & \xrightarrow{i_V} & \mathcal{R}_V & \xrightarrow{\pi_V} & Q_V & \longrightarrow & 0 . \end{array}$$

Since $\pi_U j_U = id_{Q_U}$, the mapping

$$\rho_U^V := \pi_V \; r_U^V \; j_U$$

is the unique homomorphism from Q_U into Q_V such that

$$\begin{array}{ccc} \mathcal{R}_U & \xrightarrow{\pi_U} & Q_U \\ \downarrow r_U^V & & \downarrow \rho_U^V \\ \mathcal{R}_V & \xrightarrow{\pi_V} & Q_V \end{array}$$

is commutative as well (see [9], 1.12). $\{\mathcal{R}_U, r_U^V|\mathcal{R}_U\}$ being a presheaf $\{Q_U, \rho_U^V\}$ is a presheaf. Let Q be the corresponding sheaf. The canonical mapping $\rho_U: Q_U \longrightarrow \Gamma(U,Q)$ is injective for every open set U in X. To see this suppose that $\{U_i\}_{i \in I}$ is an open covering of U and $q \in Q_U$ such that $\rho_U^{U_i} q = 0$ for all $i \in I$. Then q is harmonic on every U_i, hence harmonic on U and therefore $q = 0$.

Denoting by i the sheaf homomorphism from $\mathcal{H}$ into $\mathcal{R}$ which is induced by $\{i_U\}$ and denoting by π the sheaf homomorphism from $\mathcal{R}$ into Q which is induced by $\{\pi_U\}$ we obtain by [9], 5.8 the exact sequence

$$0 \longrightarrow \mathcal{H} \xrightarrow{i} \mathcal{R} \xrightarrow{\pi} Q \longrightarrow 0.$$

The main purpose of the next two sections is to show that this is a fine resolution of $\mathcal{H}$, i.e. that the sheaves $\mathcal{R}$ and Q are fine sheaves.

3. The sheaf $\mathcal{R}$

The presheaf $\{\mathcal{R}_U, r_U^V|\mathcal{R}_U\}$ is obviously a subpresheaf of $\{\mathcal{C}_U, r_U^V|\mathcal{C}_U\}$ where $\mathcal{C}_U$ denotes the linear space of all continuous real-valued functions on U. So $\mathcal{R}$ is a subsheaf of the sheaf $\mathcal{C}$ of germs of continuous real-valued functions on X. The isomorphism $r_U|\mathcal{C}_U$ between $\mathcal{C}_U$ and $\Gamma(U,\mathcal{C})$ hence induces an isomorphism $\tilde{r}_U$ between a subspace $\tilde{\mathcal{R}}_U$ of $\mathcal{C}_U$ and $\Gamma(U,\mathcal{R})$. The following proposition gives a characterization of $\tilde{\mathcal{R}}_U$ which shows that $\mathcal{R}$ is the sheaf of germs of differences of (non-negative and bounded) continuous real-valued superharmonic (or subharmonic) functions.

PROPOSITION 3.1: For every open subset U of X and every numerical function f on U the following statements are equivalent:

1. $f \in \tilde{\mathcal{R}}_U$.

2. For every $x \in U$ there exists an open neighborhood V of x such that $V \subset U$ and $r_U^V f \in \mathcal{R}_V$.

3. For every $x \in U$ there exists an open neighborhood V of x and (non-negative and bounded) $s_1, s_2 \in \mathcal{S}_V$ such that $V \subset U$ and $r_U^V f = s_1 - s_2$.

4. For every $x \in U$ there exists an open neighborhood V of x and (non-negative and bounded) $t_1, t_2 \in -\mathcal{S}_V$ such that $V \subset U$ and $r_U^V f = t_1 - t_2$.

PROOF: The equivalence of (1) and (2) follows immediately from the definition of $\mathcal{R}$ and $\tilde{\mathcal{R}}_U$. The equivalences of (2), (3) and (4) are based on the fact that for every $W \in \mathfrak{U}$ there exists a bounded $h_0 \in \mathcal{H}_W$ with $\inf h_0(W) > 0$, e.g. $h_0 = r_X^W H_W 1$. Assume for instance that (2) holds and take $x \in U$. Then there exist a neighborhood V of x and functions $h \in \mathcal{H}_V$, $p, q \in \mathcal{P}_V$ such that $V \subset U$ and

$$r_U^V f = h + (p - q).$$

Choose $W \in \mathfrak{U}$ such that $x \in W$ and $\overline{W} \subset V$. Since h is bounded on the compact set $\overline{W}$, there exists a bounded $h_o \in \mathcal{H}_W^+$ with $h_o \geq - r_V^W h$. Defining

$$s_1 = r_V^W(h+p) + h_o ,$$

$$s_2 = r_V^W q + h_o$$

we have

$$s_1, s_2 \in \mathcal{S}_W^+ \cap \mathcal{B}_W \qquad \text{and}$$

$$r_U^W f = s_1 - s_2 .$$

So (3) holds. The other implications follow similarly.

<u>REMARK</u>: Observing that $|s_1 - s_2| = s_1 + s_2 - 2 \inf(s_1, s_2)$ it follows that $f \in \tilde{\mathcal{R}}_U$ implies $|f| \in \tilde{\mathcal{R}}_U$.

For every open subset U of X let $\mathcal{T}_U^+$ be the set of all non-negative real-valued continuous subharmonic functions on U, i.e. $\mathcal{T}_U^+ = (-\mathcal{S}_U)^+$.

<u>LEMMA 3.2</u>: Let U be an open subset of X, $s \in \mathcal{S}_U$ and strictly positive, $t \in \mathcal{T}_U^+$ and n a natural number. Then

$$\frac{t^n}{s^{n-1}} \in \mathcal{T}_U^+ .$$

<u>PROOF</u>: Obviously $\frac{t^n}{s^{n-1}} \in \mathcal{C}_U^+$. Consider $V \in \mathfrak{U}$ such that $\overline{V} \subset U$, take an $x \in U$ and define the measure μ by

$$\int f \, d\mu = \frac{1}{s(x)} H_V(fs)(x) \qquad (f \in \mathcal{B}_X) .$$

We then have

$$(1) \quad \int 1 \, d\mu = \frac{1}{s(x)} H_V s(x) \leq 1$$

and

$$(2) \quad \int \frac{t}{s} \, d\mu = \frac{1}{s(x)} H_V t(x) \geq \frac{t}{s}(x).$$

The Hoelder inequality and (1) imply

$$(3) \quad \left(\int \frac{t}{s} \, d\mu\right)^n \leq \int \left(\frac{t}{s}\right)^n d\mu .$$

But we have

$$\int \left(\frac{t}{s}\right)^n d\mu = \frac{1}{s(x)} H_V \frac{t^n}{s^{n-1}} (x).$$

Thus (2) and (3) imply that

$$\left(\frac{t}{s}\right)^n (x) \leq \frac{1}{s(x)} H_V \frac{t^n}{s^{n-1}} (x) ,$$

i.e.
$$\frac{t^n}{s^{n-1}} (x) \leq H_V \frac{t^n}{s^{n-1}} (x).$$

So $\frac{t^n}{s^{n-1}}$ is subharmonic on U.

<u>PROPOSITION 3.3</u>: Let U be an open subset of X and $f_1, f_2, f \in \tilde{\mathcal{R}}_U$ with strictly positive f. Then $\frac{f_1 f_2}{f} \in \tilde{\mathcal{R}}_U$.

<u>PROOF</u>: Let $x \in U$. For every open subset V of U we define

$$\mathcal{R}'_V = \mathcal{T}^+_V - \mathcal{T}^+_V .$$

If W and V are open and $W \subset V \subset U$, then $r^W_V \mathcal{T}^+_V \subset \mathcal{T}^+_W$ and hence $r^W_V(\mathcal{R}'_V) \subset \mathcal{R}'_W$. By proposition 3.1 we may choose a neighborhood $V \in \mathfrak{U}$ of x in U such that $r^V_U f_1, r^V_U f_2, r^V_U f \in \mathcal{R}'_V$. We take $t_1, t_2 \in \mathcal{T}^+_V$ with

$$r^V_U f = t_1 - t_2,$$

$h_o \in \mathcal{H}_V$ with $h_o(x) = 1$ and define

$$\alpha = \frac{2}{3} t_1(x) + \frac{1}{3} t_2(x) .$$

Then

$$t_1(x) - \alpha h_o(x) = \frac{1}{3} f(x),$$

$$\alpha h_o(x) - t_2(x) = \frac{2}{3} f(x).$$

Hence there exists a neighborhood W of x in V such that

$$t: = r^W_V (t_1 - \alpha h_o) \in \mathcal{T}^+_W ,$$

$$s: = r^W_V (\alpha h_o - t_2) \in \mathcal{T}^+_W$$

and $0 < t < s$ on W. We have $s + t = r^W_U f \in \mathcal{R}'_W$. Now we may write

$$\frac{s^2}{s+t} = \frac{s}{1+\frac{t}{s}} = s+t + \sum_{n \text{ even}} \frac{t^n}{s^{n-1}} - \sum_{n \text{ odd } \geq 3} \frac{t^n}{s^{n-1}}$$

where all $\frac{t^n}{s^{n-1}} \in \mathcal{T}_W^+$ by lemma 3.2. Since $\mathcal{T}_W^+$ is closed with respect to locally uniform convergence, both countable sums in the inequality above are in $\mathcal{T}_W^+$ and hence

$$\frac{s^2}{s+t} \in \mathcal{R}'_W.$$

By lemma 3.2, we have $\frac{t'^2}{s} \in \mathcal{T}_W^+$ for all $t' \in \mathcal{T}_W^+$ and hence

$$\frac{(t_1'-t_2')^2}{s} = 2\frac{t_1'^2}{s} + 2\frac{t_2'^2}{s} - \frac{(t_1'+t_2')^2}{s} \in \mathcal{R}'_W$$

for all $t_1', t_2' \in \mathcal{T}_W^+$. Thus for all $g_1, g_2 \in \mathcal{R}'_W$

$$\frac{g_1 g_2}{s} = \frac{1}{2}\left(\frac{(g_1+g_2)^2}{s} - \frac{g_1^2}{s} - \frac{g_2^2}{s}\right) \in \mathcal{R}'_W .$$

Applying this result twice we obtain that for all $g_1, g_2 \in \mathcal{R}'_W$

$$\frac{g_1 g_2}{s+t} = \frac{\frac{g_1 g_2}{s} \cdot \frac{s^2}{s+t}}{s} \in \mathcal{R}'_W .$$

Since $r_U^W f_1$, $r_U^W f_2 \in \mathcal{R}'_W$ and $r_U^W f = s + t$ we get $r_U^W \left(\frac{f_1 f_2}{f}\right) \in \mathcal{R}'_W$. By proposition 3.1, we finally conclude that $\frac{f_1 f_2}{f} \in \tilde{\mathcal{R}}_U$.

<u>LEMMA 3.4</u>: Let $U \in \mathfrak{U}$ and K be a compact subset of U. Then there is a function $f \in \mathcal{R}_U^+$ such that $f > 0$ on K and the support $S(f) := \overline{[f \neq 0]}$ is contained in U.

<u>PROOF</u>: Because of property (III) U is a strong-harmonic subspace of X. For every $x \in U$ there hence exist $p_x \in \mathcal{P}_U$ and $V_x \in \mathfrak{U}$ such that $x \in V_x$, $\overline{V}_x \subset U$ and $H_{V_x} p_x(x) < p_x(x)$ (see [5], 12.7). We have $f_x := p_x - H_{V_x} p_x \in \mathcal{R}_U$, $f_x(x) > 0$ and $S(f_x) \subset \overline{V}_x \subset U$. There are finitely many $x_1, \dots, x_n \in K$ such that $K \subset \bigcup_{i=1}^{n} [f_{x_i} > 0]$. Then $f := \sum_{i=1}^{n} f_{x_i}$ has the required properties.

REMARK: This lemma and the remark following proposition 3.1 imply that $\tilde{\mathcal{R}}_X$ is dense in $\mathcal{C}_X$ with respect to the topology of locally uniform convergence.

COROLLARY 3.5: $\mathcal{R}$ is a fine sheaf.

PROOF: Let $\{U_j^o\}_{j \in J}$ be a locally finite open covering of X. Then there is a locally finite refinement $\{U_i\}_{i \in I}$ of $\{U_j^o\}_{j \in J}$ consisting of sets $U_i \in \mathfrak{U}$. Let $\{V_i\}_{i \in I}$ be a shrinking of $\{U_i\}_{i \in I}$. By lemma 3.4 there are $f_i \in \tilde{\mathcal{R}}_X$ such that $f_i > 0$ on $\overline{V}_i$ and $S(f_i) \subset U_i$. Then $f := \sum_{i \in I} f_i \in \tilde{\mathcal{R}}_X$ and is strictly positive. By proposition 3.3

$$c_i^U(g) := r_X^U\left(\frac{f_i}{f}\right) g \qquad (g \in \tilde{\mathcal{R}}_U)$$

defines homomorphisms $\{c_i^U\}$ of the presheaf $\{\tilde{\mathcal{R}}_U, r_U^V | \tilde{\mathcal{R}}_U\}$ with support of c_i^U in U_i and $\sum_{i \in I} c_i^U = id_{\tilde{\mathcal{R}}_U}$. Since $\mathcal{R}$ is the corresponding sheaf, the induced sheaf homomorphisms c_i are homomorphisms of the sheaf $\mathcal{R}$, have supports in the corresponding U_i and satisfy $\sum_{i \in I} c_i = id_{\mathcal{R}}$. Hence $\mathcal{R}$ is a fine sheaf.

4. The sheaf Q

For every open subset U of X and every $p \in \mathcal{P}_U$ let C(p) be the potential-theoretic support or carrier of p, i.e. the smallest closed subset A of X such that p is harmonic on $U \setminus A$. Since the constant 0 is the only harmonic potential, $C(p) = \emptyset$ is equivalent to $p = 0$.

LEMMA 4.1: Let V,U be open subsets of X, $V \subset U$ and $p \in \mathcal{P}_U$. Then $\rho_U^V p \leq p$ and $C(\rho_U^V p) \subset C(p)$. If $C(p)$ is contained in $\complement V$, then $\rho_U^V p = 0$. If $V \in \mathfrak{U}$ and $\overline{V} \subset U$, then $\rho_U^V p = p - H_V p$.

PROOF: Since $r_U^V p \in \mathcal{S}_V^+$, we have by the Riesz decomposition theorem $r_U^V p = h + q$ where $h \in \mathcal{H}_V^+$ and $q \in \mathcal{P}_V$. Thus $\rho_U^V p = q \leq r_U^V p \leq p$. q is harmonic on $V \setminus C(p)$ as well as $r_U^V p$. Hence $C(\rho_U^V p) \subset C(p)$. $C(p) \subset \complement V$ implies $r_U^V p \in \mathcal{H}_V$ and hence $\rho_U^V p = q = 0$.

Now let $V \in \mathfrak{U}$ and $\overline{V} \subset U$. We have

$$r_U^V p = r_U^V H_V p + (p - H_V p).$$

The first term on the right side is harmonic on V. The second term is in $\mathcal{S}_V^+$ and vanishes at infinity with respect to V. Since V satisfies the boundary minimum principle ([5], 5.3) the second term has greatest harmonic minorant 0 and is hence a potential. This establishes $\rho_U^V p = p - H_V p$.

A potential kernel on an open subset U of X is a finite kernel K on X such that $K(\mathcal{B}_X^+) \subset \mathcal{P}_U$ and $C(Kf) \subset S(f)$ for all $f \in \mathcal{B}_X^+$. If K is a potential kernel on U, then $K1_{\complement U}$ is a potential on U which is harmonic on U, hence $K1_{\complement U} = 0$.

PROPOSITION 4.2: Let U be an open subset of X and $p \in \mathcal{P}_U$. Then there exists exactly one potential kernel K_p on U such that $K_p 1 = p$.

PROOF: Let d be a metric on X which is compatible with the topology. By [5], 15.8, there exist $p_i^1 \in \mathcal{P}_U$ $(i \in \mathbb{N})$ such that $p = \sum_{i=1}^{\infty} p_i^1$ and all $C(p_i^1)$ are compact. Since every compact set in X may be covered by finitely many balls of diameter less than $\frac{1}{n}$, repeated application of [5], 15.6 yields potentials $p_i^n \in \mathcal{P}_U$ $(i,n \in \mathbb{N})$, such that for all $n \in \mathbb{N}$ the diameter of $C(p_i^{n+1})$ is less than $\frac{1}{n}$ for all $i \in \mathbb{N}$ and there is a partition of $\mathbb{N}$ in finite sets J_i $(i \in \mathbb{N})$ such that

$$\sum_{j \in J_i} p_j^{n+1} = p_i^n .$$

For $f \in \mathcal{C}_X^+$ having compact support define

$$K_n f = \sum_{i=1}^{\infty} \inf(f(C(p_i^n)))p_i^n .$$

Then every $K_n f$ and $K_{n+1}f - K_n f$ is a countable sum of potentials in $\mathcal{P}_U$ having carrier contained in $S(f)$. The same statement holds for

$$Kf := \sup K_n f = K_1 f + \sum_{n=1}^{\infty} (K_{n+1}f - K_n f).$$

Defining $$K^n f := \sum_{i=1}^{\infty} \sup(f(C(p_i^n)))\, p_i^n$$

we get $K^{n+1}f \leq K^n f$ and

$$0 \leq K^n f - K_n f \leq p \sup_{i \in \mathbb{N}} (\sup(f(C(p_i^n))) - \inf(f(C(p_i^n))))$$

Since the diameter of $C(p_i^n)$ tends to 0 uniformly in i as n tends to infinity and since f is uniformly continuous, this implies

$$Kf = \lim K_n f = \lim K^n f.$$

For all $f,g \in \mathcal{C}_X^+$ having compact supports we thus get

$$\begin{aligned} K(f+g) &= \lim K_n(f+g) \geq \lim K_n f + \lim K_n g \\ &= Kf + Kg = \lim K^n f + \lim K^n g \geq \lim K^n(f+g) \\ &= K(f+g). \end{aligned}$$

So the mapping $f \longrightarrow Kf$ is additive. As it is obviously isotone and positive homogeneous, it can be extended to a kernel K on X. Let (f_m) be an increasing sequence of functions in $\mathcal{C}_X^+$ having compact supports such that $\sup f_m = 1$ and $f_m = 1$ on $\bigcup_{i=1}^{m} C(p_i^1)$. Then $\sum_{i=1}^{m} p_i^1 \leq K_n f_m \leq p$ for all $m,n \in \mathbb{N}$, hence $\sum_{i=1}^{m} p_i^1 \leq Kf_m \leq p$ for all $m \in \mathbb{N}$ yielding

$$K1 = \sup Kf_m = p.$$

If $\varphi \in \mathcal{C}_X^+$, then $K\varphi$ is the supremum of the increasing sequence $(K(\varphi f_m))$ where every $K(\varphi f_m)$ is a countable sum of potentials in $\mathcal{P}_U$ having carriers in $S(\varphi)$. Having this for $K(\|\varphi\| - \varphi)$ as well and knowing

$$K\varphi + K(\|\varphi\|-\varphi) = \|\varphi\|K1 = \|\varphi\|p \in \mathcal{P}_U$$

we conclude that $K\varphi \in \mathcal{P}_U$ and $C(K\varphi) \subset S(\varphi)$.

For every lower semicontinuous $f \in \mathcal{B}_X^+$ there are $\varphi_m \in \mathcal{C}_X^+$ such that $\varphi_m \uparrow f$ showing that $Kf = \sup K\varphi_m$ is hyperharmonic. Therefore

$$Kg = \inf \{Kf : f \in \mathcal{B}_X^+ \text{ l.s.c.}, \quad f \geq g\}$$

is nearly hyperharmonic for every $g \in \mathcal{B}_X^+$.

Consider now $g \in \mathcal{B}_X^+$. Then there are $\psi_m \in \mathcal{C}_X^+$ such that $\psi_m \geq g$ and $\bigcap_{m=1}^{\infty} S(\psi_m) = S(g)$. For all $m \in \mathbb{N}$ we have

$$Kg + K(\psi_m - g) = K\psi_m \in \mathcal{P}_U$$

where $C(K\psi_m) \subset S(\psi_m)$, hence $Kg \in \mathcal{P}_U$ and $C(Kg) \subset \bigcap_{m=1}^{\infty} S(\psi_m) = S(g)$. So K is the desired kernel K_p, if we can show the uniqueness.

To this end let $\tilde{K}$ be a potential kernel on U, $\tilde{K}1 = p$ and V an open subset of U. We show

(*) $$\tilde{K}1_V = \sup\{p' \in \mathcal{P}_U : C(p') \subset V, \; p-p' \in \mathcal{P}_U\}.$$

We have

$$\tilde{K}1_V = \sup\{\tilde{K}1_L : L \text{ compact} \subset V\}.$$

But $\tilde{K}1_L \in \mathcal{P}_U$, $C(\tilde{K}1_L) \subset L \subset V$ and $p - \tilde{K}1_L = \tilde{K}1_{\complement L} \in \mathcal{P}_U$ for all compact subsets L of V. Hence the left side of (*) is less or equal to the right side. Consider now $p' \in \mathcal{P}_U$ such that $C(p') \subset V$ and $p - p' \in \mathcal{P}_U$. By [5], 15.4 there are $p_1, p_2 \in \mathcal{P}_U$ such that $p' = p_1 + p_2$, $\tilde{K}1_V - p_1 \in \mathcal{P}_U$, $\tilde{K}1_{\complement V} - p_2 \in \mathcal{P}_U$. Then $C(p_2) \subset C(\tilde{K}1_{\complement V}) \cap C(p') \subset \complement V \cap V = \emptyset$, i.e. $p_2 = 0$. Thus $p' = p_1 \leq \tilde{K}1_V$. This proves the rest.

Let U be an open subset of X, $p_1, p_2 \in \mathcal{P}_U$ and $\alpha_1, \alpha_2 \geq 0$. Then $\alpha_1 K_{p_1} + \alpha_2 K_{p_2}$ is a potential kernel on U satisfying $(\alpha_1 K_{p_1} + \alpha_2 K_{p_2})(1) = \alpha_1 p_1 + \alpha_2 p_2$. Hence by proposition 4.2,

$$K_{\alpha_1 p_1 + \alpha_2 p_2} = \alpha_1 K_{p_1} + \alpha_2 K_{p_2}.$$

Therefore the mapping $p \longrightarrow K_p$ from $\mathcal{P}_U$ into the space $\mathcal{L}_U$ of all linear mappings from $\mathcal{B}_X$ into Q_U can be extended in a unique way to a linear mapping $q \longrightarrow K_q$ from Q_U into $\mathcal{L}_U$, namely by

$$K_{p_1 - p_2} = K_{p_1} - K_{p_2}.$$

<u>LEMMA 4.3</u>: For all $q \in Q_U$ and $f, g \in \mathcal{B}_X$

$$K_{K_q f} g = K_q(fg).$$

<u>PROOF</u>: Because of the linearity it suffices to consider $q \in \mathcal{P}_U$ and $f \in \mathcal{B}_X^+$. Define K by

$$Kg = K_q(fg) \qquad (g \in \mathcal{B}_X).$$

Obviously K is a potential kernel on U and $K1 = K_q f$. Hence $K = K_{K_q f}$ which establishes the lemma.

<u>COROLLARY 4.4:</u> Q_U is a $\mathcal{B}_U$ - module with respect to the multiplication $f \circ q := K_q f$.

<u>LEMMA 4.5</u>: Let U, V be open subsets of X, $V \subset U$. Then for all $q \in Q_U$ and $f \in \mathcal{B}_X$

$$\rho_U^V (K_q f) = K_{\rho_U^V q} f .$$

<u>PROOF</u>: Again it is sufficient to consider $q \in \mathcal{P}_U$. Define K by

$$Kf = \rho_U^V (K_q f) \qquad (f \in \mathcal{B}_X).$$

Then K is a linear mapping from $\mathcal{B}_X$ into Q_V such that $K(\mathcal{B}_X^+) \subset \mathcal{P}_V$. By lemma 4.1, we have $C(Kf) \subset C(K_q f) \subset S(f)$ for all $f \in \mathcal{B}_X^+$ and $K \leq K_q$. Hence K is a potential kernel on V. Thus

$$K1 = \rho_U^V (K_q 1) = \rho_U^V q$$

implies $K = K_{\rho_U^V q}$ which establishes the lemma.

COROLLARY 4.6: $\{\mathfrak{B}_U, Q_U, \{r_U^V, \rho_U^V\}\}$ is a presheaf

PROOF: Let V, U be open subsets of X, $V \subset U$, $f \in \mathfrak{B}_U$ and $q \in Q_U$. Then by lemma 4.5

$$\rho_U^V(f \circ q) = \rho_U^V(K_q f) = K_{\rho_U^V q} f = K_{\rho_U^V q}(r_U^V f) = (r_U^V f) \circ (\rho_U^V q) .$$

This and corollary 4.4 prove the statement.

Denoting by $\mathfrak{B}$ the sheaf of germs of bounded Borel measurable functions we obtain the

PROPOSITION:4.7: Q is a $\mathfrak{B}$-module.

COROLLARY 4.8: Q is a fine sheaf.

PROOF: Let A, B be disjoint closed subsets of X. There is an open neighborhood U of A such that $\overline{U} \cap B = \emptyset$. We define $c: Q \longrightarrow Q$ by

$$c(q^x) = (r_X^x 1_U) \circ q^x \qquad (x \in X,\ q^x \in Q_x).$$

c is a homomorphism, $c(q^x) = q^x$, if $x \in U$, and $c(q^x) = 0_x$, if $x \in \complement \overline{U}$. Hence Q is a fine sheaf by [9], 14.4.

REMARK: Since Q is a $\mathfrak{B}$-module every $\Gamma(U,Q)$ (U open in X) becomes a $\mathfrak{B}_U$-module by the definition

$$(f \circ M)(x) = (r_U^x f) \circ (M(x)) \qquad (f \in \mathfrak{B}_U,\ M \in \Gamma(U,Q),\ x \in U).$$

We note that $(f \circ M)(x) = 0_x$ for all $x \in U \setminus S(f)$.

Every Q_U is ordered by the pointed cone $Q_U^+ := \mathcal{P}_U$. We have $\rho_U^V \mathcal{P}_U \subset \mathcal{P}_V$ for all open subsets V of U. For every $x \in X$ the linear space Q_x is hence ordered by the pointed cone

$$Q_x^+ = \bigcup_{U \text{ open},\ x \in U} \rho_U^x \mathcal{P}_U$$

and

$$Q = Q_x^+ - Q_x^+ .$$

The corresponding order will be denoted by $\prec$. For every open U in X let

$$\Gamma^+(U,Q) = \{M \in \Gamma(U,Q) : M(x) \in Q_x^+ \text{ for all } x \in U\}.$$

PROPOSITION 4.9: Let U be an open subset of X. Then

1. $f \circ M \in \Gamma^+(U,Q)$ for all $f \in \mathcal{B}_U^+$ and $M \in \Gamma^+(U,Q)$.
2. $\Gamma(U,Q) = \Gamma^+(U,Q) - \Gamma^+(U,Q)$.

PROOF: (1) follows from the fact that $g \circ p \in \mathcal{P}_V$ for all open V in U, $g \in \mathcal{B}_V^+$ and $p \in \mathcal{P}_V$.

(2) Consider $M \in \Gamma(U,Q)$. There is an open locally finite covering $\{U_i\}_{i \in I}$ such that for every $i \in I$ there are potentials $p_i, q_i \in \mathcal{P}_{U_i}$ satisfying

$$M \mid U_i = \rho_{U_i} (p_i - q_i) .$$

Let $\{V_i\}_{i \in I}$ be a shrinking of $\{U_i\}_{i \in I}$. Then for every $i \in I$

$$M_i := \rho_{U_i}(1_{V_i} \circ p_i) \in \Gamma^+(U_i,Q)$$

having support in $\overline{V}_i$. Using the value zero we hence may extend M_i to a section $N_i \in \Gamma^+(U,Q)$. Define

$$N = \sum_{i \in I} N_i .$$

Then $N \in \Gamma^+(U,Q)$ and for all $i \in I$

$$N|V_i \succ N_i|V_i = M_i|V_i = \rho_{V_i} \rho_{U_i}^{V_i} (1_{V_i} \circ p_i)$$

$$= \rho_{V_i} \rho_{U_i}^{V_i} p_i \succ \rho_{V_i} \rho_{U_i}^{V_i} (p_i - q_i) = M|V_i .$$

Since $\{V_i\}_{i \in I}$ is a covering of U, this shows $N - M \in \Gamma^+(U,Q)$ and we obtain

$$M = N - (N-M) \in \Gamma^+(U,Q) - \Gamma^+(U,Q).$$

PROPOSITION 4.10: Let U be a strong-harmonic subspace of X and $M \in \Gamma(U,Q)$ having compact support in U. Then there exists exactly one $q \in Q_U$ such that $\rho_U q = M$. If in addition $M \succ 0$, then $q \in \mathcal{P}_U$.

PROOF: For every point x in the support K of M there is an open relatively compact set U_x and $p_x, q_x \in \mathcal{P}_{V_x}$ such that $x \in U_x \subset U$ and

$$M|U_x = \rho_{U_x}(p_x - q_x).$$

Choose open V_x such that $x \in V_x$, $\overline{V}_x \subset U_x$. Then there are finitely many $x_1, \dots, x_n \in K$ such that $K \subset \bigcup_{i=1}^{n} V_i$ where the subscript x_i is replaced by i as will be done in the following. For $1 \leq i \leq n$ define

$$Z_i = V_i \setminus \bigcup_{j=1}^{i-1} V_j \quad , \quad Z = \bigcup_{i=1}^{n} Z_i = \bigcup_{i=1}^{n} V_i \ .$$

Since the support K of M is contained in Z, we have

$$M = 1_Z \circ M = \sum_{i=1}^{n} 1_{Z_i} \circ M$$

(see the remark following corollary 4.8). Let $1 \leq i \leq n$. The support of $1_{Z_i} \circ M$ is contained in $\overline{Z}_i$, hence in U_i and we get for all $x \in U_i$

$$(1_{Z_i} \circ M)(x) = (r^x_{U_i} 1_{Z_i}) \circ (M(x))$$

$$= (r^x_{U_i} 1_{Z_i}) \circ (\rho^x_{U_i}(p_i - q_i)) = \rho^x_{U_i}(1_{Z_i} \circ p_i - 1_{Z_i} \circ q_i).$$

Now $p_i' := 1_{Z_i} \circ p_i \in \mathcal{P}_{U_i}$ and $C(p_i') \subset \overline{Z}_i \subset U_i$. By the extension theorem of M. Hervé ([7], 13.2) which because of [5], 16.1 holds in our situation as well there exists a potential $\tilde{p}_i \in \mathcal{P}_U$ such that $C(\tilde{p}_i) = C(p_i')$ and $r_U^{U_i} \tilde{p}_i - p_i' \in \mathcal{H}_{U_i}$. This implies $\rho_U^{U_i} \tilde{p}_i = p_i'$. By the same reason there is a $\tilde{q}_i \in \mathcal{P}_U$ such that $\rho_U^{U_i} \tilde{q}_i = q_i' := 1_{Z_i} \circ q_i$ and $C(\tilde{q}_i) = C(q_i')$. We obtain for all $x \in U_i$

$$(1_{Z_i} \circ M)(x) = \rho^x_{U_i}(p_i' - q_i') = \rho^x_U(\tilde{p}_i - \tilde{q}_i) \ .$$

On the other hand $\tilde{p}_i$ and $\tilde{q}_i$ are harmonic on $U \setminus \overline{Z}_i$. Hence we have for all $x \in U \setminus U_i \subset U \setminus \overline{Z}_i$

$$(1_{Z_i} \circ M)(x) = 0_x = \rho_U^x(\tilde{p}_i - \tilde{q}_i),$$

So $q := \sum_{i=1}^{n} (\tilde{p}_i - \tilde{q}_i) \in Q_U$ and $\rho_U q = M$. On p. 69 we saw that ρ_U is injective. Therefore q is uniquely determined.

If in addition $M \succ 0$, we may choose $q_x = 0$ for sufficiently small U_x leading to $q \in \mathcal{P}_U$.

PROPOSITION 4.11: Let $M \in \Gamma^+(X,Q)$ and (f_n) be a decreasing sequence in $\mathcal{B}_X^+$ such that $\inf f_n = 0$. Then the zero section is the infimum of $(f_n \circ M)$ in the order of $\Gamma^+(X,Q)$.

PROOF: Let $N \in \Gamma^+(X,Q)$ and $N \prec f_n \circ M$ for all $n \in \mathbb{N}$. We have to show $N = 0$.

Consider first a set $U \in \mathfrak{U}$ and suppose that the supports of all sections in question are compact subsets of U (which already holds if the support of M is a compact subset of U). By proposition 4.10, there are uniquely determined $q, p, p_n \in \mathcal{P}_U$ such that

$$N|U = \rho_U q, \quad M|U = \rho_U p, \quad (f_n \circ M - N) \mid U = \rho_U p_n.$$

But we know

$$f_n \circ M|U = \rho_U(f_n \circ p)$$

from the definition of $f_n \circ M$. Hence $N + (f_n \circ M - N) = f_n \circ M$ implies

$$q + p_n = f_n \circ p = K_p f_n.$$

So we conclude that $0 \leq q \leq K_p f_n$ for all n. But $\inf K_p f_n = 0$, since $f_n \downarrow 0$ and K_p is a kernel. Hence $q = 0$, i.e. $N|U = 0$ and therefore $N = 0$.

Consider now the general case. There is a locally finite covering $\{U_m\}_{m \in \mathbb{N}}$ of X consisting of sets in $\mathfrak{U}$. Let $\{V_m\}_{m \in \mathbb{N}}$ be a shrinking of $\{U_m\}_{m \in \mathbb{N}}$ and define $\{Z_m\}_{m \in \mathbb{N}}$ by

$$Z_m = V_m \setminus \bigcup_{i=1}^{m-1} V_i.$$

Then $1_{Z_m} \circ N < 1_{Z_m} \circ (f_n \circ M) = f_n \circ (1_{Z_m} \circ M)$ for all $n, m \in \mathbb{N}$. Hence our reasoning above shows $1_{Z_m} \circ N = 0$ for all $m \in \mathbb{N}$ giving

$$N = \sum_{m=1}^{\infty} 1_{Z_m} \circ N = 0.$$

5. Cohomology groups of $\mathcal{H}$

Since $0 \longrightarrow \mathcal{H} \xrightarrow{i} \mathcal{R} \xrightarrow{\pi} Q \longrightarrow 0$ is a fine resolution of $\mathcal{H}$, the cohomology groups of $\mathcal{H}$ are just the cohomology groups of the cochain complex

$$0 \longrightarrow \Gamma(X,\mathcal{R}) \xrightarrow{d} \Gamma(X,Q) \longrightarrow 0 \longrightarrow \ldots$$

where d is the mapping induced by π, i.e. $d(\sigma) = \pi\sigma$ ([9], 14.6, 18.8, 18.10). Therefore

$$H^0(X,\mathcal{H}) \cong \ker d$$

$$H^1(X,\mathcal{H}) \cong \operatorname{coker} d = \Gamma(X,Q) / d\Gamma(X,\mathcal{R})$$

and we have the

THEOREM 5.1: $H^q(X,\mathcal{H}) = 0$ for all $q \geq 2$.

There are immediate examples of non-compact harmonic spaces for which $\dim \mathcal{H}_X = \infty$ and hence $\dim H^0(X,\mathcal{H}) = \infty$, since $H^0(X,\mathcal{H}) \cong \Gamma(X,\mathcal{H}) \cong \mathcal{H}_X$. (Consider the solutions of the Laplace equation in an open disc.) But we have

PROPOSITION 5.2: If X is compact, then $\dim H^0(X,\mathcal{H}) < \infty$.

PROOF: Suppose that X is compact. Given the supremum norm $\mathcal{H}_X$ is a Banach space. Its unit ball is relatively compact, since for all $U \in \mathfrak{U}$ the function $H_U f$ is continuous on U for every $f \in \mathcal{B}_X$ and $H_U h = h$ for all $h \in \mathcal{H}_X$ (see [10], p. 172). Being a locally compact normed space $\mathcal{H}_X$ is thus finite-dimensional. This proves the proposition, since $H^0(X,\mathcal{H}) \cong \mathcal{H}_X$.

In the next section we shall see how every harmonic

space can be turned into a strong harmonic space by a sufficient perturbation. Using that we shall show that for every compact harmonic space X the dimension of $H^1(X,\mathcal{H})$ is finite as well and in fact equal to the dimension of $H^o(X,\mathcal{H})$. Obviously this amounts to proving that d is a Fredholm operator of index 0. However the case of a strong harmonic X can be treated right away:

THEOREM 5.3: Let X be strong harmonic and compact. Then $H^o(X,\mathcal{H}) = H^1(X,\mathcal{H}) = 0$, i.e. d is an isomorphism. Furthermore $\Gamma(X,Q) = \rho_X Q_X$ and $\Gamma(X,\mathcal{R}) = r_X Q_X$.

PROOF: We know $\mathcal{H}_X = 0$ by [5], 13.2, hence $H^o(X,\mathcal{H}) = 0$ and d is injective. We have $\Gamma(X,Q) = \rho_X Q_X$ by proposition 4.10. But for every $q \in Q_X$ $r_X q \in \Gamma(X,\mathcal{R})$ and $d(r_X q) = \pi r_X q = \rho_X q$. Hence d is surjective and $H^1(X,\mathcal{H}) = 0$. Furthermore $\Gamma(X,\mathcal{R}) = r_X Q_X$, since d is injective.

A supplementary result is the following

PROPOSITION 5.4: Suppose that X is compact, connected and elliptic and that there is a non-negative superharmonic function s on X which is not identically 0. If s is harmonic, then $\dim \mathcal{H}_X = 1$. If not, then $\dim \mathcal{H}_X = 0$.

PROOF: We note first that s is strictly positive by [5], 6.3. Consider $h \in \mathcal{H}_X$. There is an $\alpha > 0$ such that $h \leq \alpha s$. Choosing

$$\beta := \inf \{\alpha \in \mathbb{R} : h \leq \alpha s\}$$

$\beta s - h$ is a non-negative superharmonic function which is zero at least at one point. By [5], 6.3 again, we conclude that $\beta s - h = 0$ i.e. $h = \beta s$. Now the two statements follow easily.

It is clear that the compactness and connectedness of X is essential for the statement $\dim \mathcal{H}_X \leq 1$. Using the same underlying topological space the following two examples show that the ellipticity and existence of a non-trivial non-negative super-harmonic function are crucial as well.

Let $e_1 = (1,0)$ and $e_2 = (0,1)$ be the unit vectors in $\mathbb{R}^2$, n an arbitrary natural number,

$$m_i = i\, e_1$$
$$a_i = m_i - e_2 \qquad (1 \leq i \leq n)$$
$$b_i = m_i + e_2$$

and

$$X = [m_1, m_n] \cup \bigcup_{i=1}^{n} [a_i, b_i]$$

provided with the metric d induced by the euclidean $\mathbb{R}^2$:

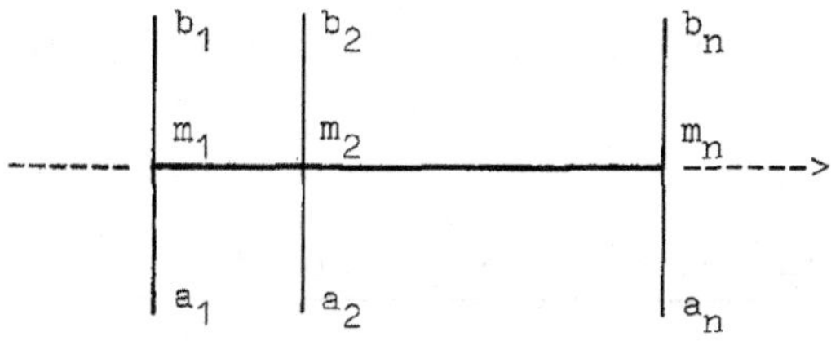

Let

$$M = \{m_i : 1 \leq i \leq n\}, \quad E = \bigcup_{i=1}^{n} \{a_i, b_i\}.$$

Let $\mathfrak{U}$ be the family of all open connected subsets of X having diameter < 1. For all $U \in \mathfrak{U}$ and $f \in \mathcal{B}_X$ define $H_U f = f$ on $\complement U$. We consider the following two definitions:

1) On $\overline{U}$ let $H_U f$ be affine-linear, if $U \cap (M \cup E) = \emptyset$, constant, if $U \cap E \neq \emptyset$, and affine linear on $U \cap [a_i, b_i]$, $U \cap [m_{i-1}, m_i]$ and $U \cap [m_i, m_{i+1}]$, if $m_i \in U$.

It is easily seen that $\{H_U\}_{U \in \mathfrak{U}}$ then is a family of local harmonic kernels on X such that $1 \in {}^*\mathcal{H}^+$ (but not elliptic, if $n \geq 2$: every $[a_i, b_i]$ is an absorbing set). Obviously $\mathcal{H}_X$ is the set of all real-valued functions on X which have an arbitrary constant value on every interval $[a_i,b_i]$ and are affine-linear on every interval $[m_i,m_{i+1}]$. Hence $\dim \mathcal{H}_X = n$.

2. Now define $H_U f$ on $\overline{U}$ in the following way: Again affine-linear, if $U \cap (M \cup E) = \emptyset$, but a multiple of the function $x \longrightarrow d(x,m_i)$, if $U \cap \{a_i,b_i\} \neq \emptyset$. If $m_i \in U$, let $H_U f$ be affine-linear on $U \cap [a_i,m_i]$, $U \cap [m_i,b_i]$, $U \cap [m_{i-1},m_i]$, $U \cap [m_i,m_{i+1}]$ and

$$H_U f(m_i) = \sum_{z \in U^*} d(z,m_i) \sum_{z \in U^*} \frac{f(z)}{d(z,m_i)} .$$

Again $\{H_U\}_{U \in \mathfrak{U}}$ is a family of local harmonic kernels, now elliptic. $\mathcal{H}_X$ is the set of all real-valued functions on X which on every $[a_i,b_i]$ are multiples of the function $m_i + \lambda e_2 \longrightarrow \lambda$ and are zero on $[m_1,m_n]$. Hence $\dim \mathcal{H}_X = n$.

REMARK: For every natural number n there exists a family of harmonic kernels on the closed unit square such that $\dim \mathcal{H}_X = n$.

6. Perturbation of the harmonic structure

We will now develop a perturbation of the harmonic space which generalizes the change from $\Delta u = 0$ to $\tilde{\Delta} u = cu$ in the classical situation. To see how this can be done suppose for a moment that $X = \mathbb{R}^n$, the harmonic functions are the solutions of $\Delta u = 0$ and $c \in \mathcal{C}_X^+$. Consider a regular set U in $\mathbb{R}^n$ which is so small that $\|G(r_X^U c)\| < 1$ where G denotes the Green kernel on U. Let K_U^c be the kernel $f \longrightarrow G(r_X^U(cf))$ on X. Consider $f \in \mathcal{C}_X$, the Dirichlet solution $u = H_U f$ and define

$$v = (I + K_U^c)^{-1} u .$$

Then $v = u - K_U^c v$ and we have on U

$$\tilde{\Delta} v = \tilde{\Delta} u - \tilde{\Delta} G(r_X^U(cv)) = \varkappa_n cv$$

where $\tilde{\Delta}$ is the generalized Laplacian and $\varkappa_n$ is a constant > 0 ([6]. 6.25). Furthermore $v = u = f$ on $\complement U$. So v is the

Dirichlet solution for U, f and $\tilde{\Delta} v = \varkappa_n cv$.
Let us finally note that of course K_U^c is the potential kernel $K_{G(r_X^U c)}$ on U and that c defines a section $M \in \Gamma^+(X,Q)$:
$M|U = \rho_U(G(r_X^U c))$.

Returning to the general situation we shall hence take $M \in \Gamma^+(X,Q)$ and consider the operators $(I + K_U^M)^{-1} H_U$ where $K_U^M = K_p$ if $M|U = \rho_U p$ with $p \in \mathcal{P}_U$ tending to zero at U^* and $\|p\| < 1$. We shall see that these operators form a family of local harmonic kernels on X which is strong-harmonic if M is properly chosen.

For $U \in \mathfrak{U}$ let $\mathcal{P}_U^1$ denote the set of all $p \in \mathcal{P}_U$ having a supremum norm $\|p\|$ strictly less than 1. We have the important

LEMMA 6.1: Let $U \in \mathfrak{U}$. We then have:

1. For every $p \in \mathcal{P}_U^1$ the operator $I + K_p$ on $\mathcal{B}_X$ has an inverse, namely

$$(I+K_p)^{-1} = I + \sum_{n=1}^{\infty} (-K_p)^n.$$

2. If $p \in \mathcal{P}_U^1$ and s is a bounded function in $\mathcal{S}_U^+$, then $(I+K_p)^{-1}s \geq 0$ with strict inequality on $[s > 0]$.

3. If $p, q \in \mathcal{P}_U^1$ with $p < q$ and s is a bounded function in $\mathcal{S}_U^+$, then

$$(I+K_p)^{-1}s \geq (I+K_q)^{-1}s.$$

If in addition s is strictly positive on U, we have strict inequality on $[p < q]$.

PROOF: (1) is trivial since $\|K_p\| = \|p\| < 1$.
2. Consider $p \in \mathcal{P}_U^1$, a bounded $s \in \mathcal{S}_U^+$ and let $t = (I+K_p)^{-1}s$ and $V = [t < 0]$. Since $t = s - K_p t$, t is continuous on U, hence

V an open subset of U. $K_p t^- \in \mathcal{P}_U$ and $C(K_p t^+) \subset S(t^+) \subset \complement V$ imply that $t = s + K_p t^- - K_p t^+$ is superharmonic on V. Furthermore $t \geq -K_p t^+$ and $K_p t^+$ is a potential on U. Since t equals 0 on $V^* \cap U$, the boundary minimum principle ([5], 11.5) implies $t \geq 0$ on V, i.e. $V = \emptyset$, and hence $t \geq 0$ on U.

Before finishing the proof of (2) we shall show the first part of (3): Let $q \in \mathcal{P}_U^1$ with $p \prec q$, i.e. $q - p \in \mathcal{P}_U$. We then have

$$(I+K_p)^{-1}s - (I+K_q)^{-1}s$$
$$= (I+K_q)^{-1}((I+K_q) - (I+K_p))(I+K_p)^{-1}s$$
$$= (I+K_q)^{-1} K_{q-p} (I+K_p)^{-1}s .$$

By (2), $t = (I+K_p)^{-1}s \in \mathcal{B}_X^+$. Therefore $u := K_{q-p}t \in \mathcal{P}_U$ and $(I+K_q)^{-1}u \geq 0$ by (2) again. So we get

$$(I+K_p)^{-1}s \geq (I+K_q)^{-1}s.$$

Now there exists an $\varepsilon > 0$ such that $\|(1+\varepsilon)p\| < 1$. For every $\alpha \in [0, 1+\varepsilon]$ we have $\alpha p \in \mathcal{P}_U^1$ and for every $x \in U$

$$(I+K_{\alpha p})^{-1}s(x) = s(x) + \sum_{n=1}^{\infty} (-K_p)^n s(x)\alpha^n.$$

Consider now $x \in U$ such that $(I+K_p)^{-1}s(x) = 0$. Having $p \prec \alpha p$ for every $\alpha \in [1,1+\varepsilon]$ our considerations above show that $(I+K_{\alpha p})^{-1}s(x) = 0$ for every $\alpha \in [1,1+\varepsilon]$. Being a power series in α the right side of the equality above hence is identically zero. In particular, $s(x) = 0$. This proves the rest of (2).

Suppose now that s is strictly positive on U. Using the same notations as above we then conclude that t is strictly positive on U, hence u is strictly positive on $[p < q]$ and therefore $(I+K_p)^{-1}s > (I+K_q)^{-1}s$ on $[p < q]$.

<u>COROLLARY 6.2</u>: Let $U \in \mathfrak{U}$. We then have:

1) For every $p \in \mathcal{P}_U^1$, $G_U^p := (I+K_p)^{-1}H_U$

is a strictly positive boundary kernel for U.

2) If $p, q \in \mathcal{P}_U^1$ and $q \prec p$, then $G_U^p \leq G_U^q$. In particular $G_U^p \leq H_U$.

PROOF: Let $p \in \mathcal{P}_U^1$. Then G_U^p is a linear operator on $\mathcal{B}_X$. Consider $f \in \mathcal{B}_X^+$ and $x \in X$. If $x \in \complement U$, then $K_p g(x) = 0$ for all $g \in \mathcal{B}_X$, hence in particular

$$G_U^p f(x) = H_U f(x) = f(x).$$

If $x \in U$, then $(I+K_p)^{-1}(1_{\complement U} g)(x) = (1_{\complement U} g)(x) = 0$ for all $g \in \mathcal{B}_X$, taking $g = H_U f$ we hence obtain

$$G_U^p f(x) = (I+K_p)^{-1}(1_U H_U f)(x).$$

Since $1_U H_U f$ is a bounded function in $\mathcal{H}_U^+$, we have $G_U^p f \geq 0$ on U by lemma 6.1,2. So G_U^p is a positive operator. Since $H_U 1$ is strictly positive, the same lemma implies that G_U^p is strictly positive as well. Because $1_U H_U 1_{\complement U^*} = 0$ we have $G_U^p 1_{\complement U^*}(x) = 0$ for all $x \in U$.

Consider now $q \in \mathcal{P}_U$ with $q \prec p$. By lemma 6.2,3 we have for every $x \in U$

$$G_U^p f(x) = (I+K_p)^{-1}(1_U H_U f)(x)$$
$$\leq (I+K_q)^{-1}(1_U H_U f)(x) = G_U^q f(x),$$

hence $G_U^p \leq G_U^q$. In particular, $G_U^p \leq G_U^o = H_U$. This finally shows that G_U^p is a kernel on X.

For $U \in \mathfrak{U}$ we define $\mathcal{P}_U^o := \mathcal{P}_U^1 \cap \mathcal{C}_X$. So $\mathcal{P}_U^o$ is the set of all $p \in \mathcal{P}_U^1$ satisfying $\lim\limits_{x \to z} p(x) = 0$ for all $z \in U^*$. For every $M \in \Gamma^+(X,Q)$ let

$$\mathfrak{U}^M = \{U \in \mathfrak{U} : M|U \in \rho_U \mathcal{P}_U^o\}.$$

If $M \in \Gamma^+(X,Q)$ and $U \in \mathfrak{U}^M$, there exists a unique $p \in \mathcal{P}_U^o$ with $M|U = \rho_U p$. We then define

$$K_U^M = K_p.$$

K_U^M is a strong Feller kernel on X, i.e. $K_U^M(\mathcal{B}_X) \subset \mathcal{C}_X$: For every $f \in \mathcal{B}_X$ we have $K_p f \in Q_U$ and $|K_p f| \leq \|f\| p$, hence $\lim_{x \to z} K_p f(x) = 0$ for every $z \in U^*$.

LEMMA 6.3: Let $M \in \Gamma^+(X,Q)$, $U \in \mathfrak{U}^M$ and $V \in \mathfrak{U}$ with $\overline{V} \subset U$. Then $V \in \mathfrak{U}^M$ and

$$K_V^M = K_U^M - H_V K_U^M .$$

PROOF: Let $p \in \mathcal{P}_U^o$ with $M|U = \rho_U p$. Then $M|V = \rho_V \rho_U^V p$. But by lemma 4.1,

$$\rho_U^V p = p - H_V p.$$

So $\rho_U^V p \in \mathcal{P}_V^o$ and hence $V \in \mathfrak{U}^M$. By the same reason we have for every $f \in \mathcal{B}_X^+$

$$\rho_U^V(K_p f) = K_p f - H_V K_p f$$

whereas on the other hand

$$\rho_U^V(K_p f) = K_{\rho_U^V p} f$$

by lemma 4.5. Therefore we obtain

$$K_{\rho_U^V p} = K_p - H_V K_p .$$

i.e.

$$K_V^M = K_U^M - H_V K_U^M .$$

For every $M \in \Gamma^+(U,Q)$ and $U \in \mathfrak{U}^M$ we define

$$G_U^M := (I + K_U^M)^{-1} H_U .$$

PROPOSITION 6.4: Let $M \in \Gamma^+(X,Q)$. Then $\{G_U^M\}_{U \in \mathfrak{U}^M}$ is a family of local harmonic kernels on X.

PROOF: We first show that $\mathfrak{U}^M$ is a base of X. Consider $x \in X$ and a neighborhood V of x. There exists an open neighborhood U of x with $U \subset V$ and a $q \in \mathcal{P}_U$ with $M|U = \rho_U q$. Take a decreasing sequence (U_n) in $\mathfrak{U}$ such that $\overline{U}_1 \subset U$ and

$\bigcap_{n=1}^{\infty} U_n = \{x\}$. Then $(H_{U_n} q)$ is an increasing sequence in $\mathcal{P}_U$ with $\sup H_{U_n} q = q$. By Dini's theorem there exists a natural m such that $q \leq H_{U_m} q + \frac{1}{2}$. We then have $p: = q - H_{U_m} q \in \mathcal{P}^{o}_{U_m}$ and $M|U_m = \rho_{U_m} p$. Thus $U_m \in \mathfrak{U}^M$.

Now let $U \in \mathfrak{U}^M$. By corollary 6.2, G_U^M is a strictly positive boundary kernel for U satisfying $G_U^M \leq H_U$. For every $f \in \mathcal{B}_X$ we have

$$G_U^M f = H_U f - K_U^M G_U^M f.$$

Since K_U^M is a strong Feller kernel, $K_U^M G_U^M f$ is continuous on X. Hence the continuity properties (I) of $H_U f$ imply the same properties for $G_U^M f$.

For every $V \in \mathfrak{U}^M$ with $\overline{V} \subset U$ we have by lemma 6.3

$$(I + K_V^M) G_U^M = G_U^M + K_U^M G_U^M - H_V K_U^M G_U^M$$

$$= H_U - H_V K_U^M G_U^M = H_V(H_U - K_U^M G_U^M) = H_V G_U^M$$

and hence

$$G_U^M = (I + K_V^M)^{-1} H_V G_U^M = G_V^M G_U^M.$$

If $s: U \longrightarrow [0, +\infty]$ is lower semicontinuous and $H_V s \leq s$ for all $V \in \mathfrak{U}$ with $\overline{V} \subset U$, we have a fortiori $G_V^M s \leq s$ for all $V \in \mathfrak{U}^M$ with $\overline{V} \subset U$ because $G_V^M \leq H_V$. This gives the last property of a family of local harmonic kernels.

For every $M \in \Gamma^+(X,Q)$ and every open subset U of X we denote by ${}^M\mathcal{H}_U$ the set of all M-harmonic functions on U, i.e. all functions on U which are harmonic with respect to $\{G_V^M\}_{V \in \mathfrak{U}^M}$. Analogous notations for hyperharmonic, superharmonic, subharmonic functions and potentials, the corresponding presheaves and sheaves.

<u>PROPOSITION 6.5</u>: For every $M, N \in \Gamma^+(X,Q)$ with $N \prec M$ and every open subset U of X we have ${}^{N*}\mathcal{H}_U^+ \subset {}^{M*}\mathcal{H}_U^+$, ${}^N\mathcal{P}_U \subset {}^M\mathcal{P}_U$ and ${}^M\mathcal{J}_U^+ \subset {}^N\mathcal{J}_U^+$.

PROOF: Let $M, N \in \Gamma^+(X,Q)$ with $N \prec M$. By corollary 6.2, we then have $G_V^M \leq G_V^N$ for every $V \in \mathfrak{U}^M \cap \mathfrak{U}^N$. Since $\mathfrak{U}^M \cap \mathfrak{U}^N$ is a base of X by lemma 6.3, we obtain ${}^{N*}\mathcal{H}_U^+ \subset {}^{M*}\mathcal{H}_U^+$ and ${}^{M}\mathcal{T}_U^+ \subset {}^{N}\mathcal{T}_U^+$ for every open U in X.

Now consider $p \in {}^{N}\mathcal{P}_U$ and $h \in {}^{M}\mathcal{H}_U^+$ with $h \leq p$. Then $h \in {}^{M}\mathcal{T}_U^+ \subset {}^{N}\mathcal{T}_U^+$ and hence $h = 0$. Thus $p \in {}^{M}\mathcal{P}_U$.

In connection with proposition 3.1 we obtain the following

COROLLARY 6.6: ${}^{M}\mathcal{R} = \mathcal{R}$ and ${}^{M}\tilde{\mathcal{R}}_U = \tilde{\mathcal{R}}_U$ for every $M \in \Gamma^+(X,Q)$ and open subset U of X.

In the following let M be an arbitrary section of $\Gamma^+(X,Q)$, if not stated otherwise.

LEMMA 6.7: For every open subset U of X and every $f \in \mathcal{C}_U$ the following three statements are equivalent:

1. f is M-superharmonic on U.
2. $f + K_V^M f$ is superharmonic on V for every $V \in \mathfrak{U}^M$ with $\overline{V} \subset U$.
3. The family of all $V \in \mathfrak{U}^M$ with $\overline{V} \subset U$ such that $f + K_V^M f$ is superharmonic on V forms a base of U.

PROOF: Let U be open in X and $f \in \mathcal{C}_U$. Without loss of generality we may assume that f is bounded. Suppose first that f is M-superharmonic on U and consider $V \in \mathfrak{U}^M$ with $\overline{V} \subset U$. For every $W \in \mathfrak{U}^M$ with $\overline{W} \subset V$ we then have $f - G_W^M f \geq 0$, hence

$$(f + K_V^M f) - H_W(f + K_V^M f)$$
$$= f + K_W^M f - H_W f = (I + K_W^M)(f - G_W^M f) \geq 0.$$

Therefore $f + K_V^M f$ is superharmonic on V.

(2) trivially implies (3), since $\mathfrak{U}^M$ is a base of X. Assume now that $V \in \mathfrak{U}^M$ such that $\overline{V} \subset U$ and $f + K_V^M f$ is superharmonic on V.

Then $g := f + K_V^M f - H_V f$ is superharmonic on V as well. But g is continuous on U and zero on $\complement V$. Hence $g \geq 0$ by the boundary minimum principle. So lemma 6.1 yields

$$f - G_V^M f = (I + K_V^M)^{-1} g \geq 0.$$

Thus (3) implies (1).

Since Q is a $\mathcal{B}$-sheaf and $\mathcal{R}$ is a subsheaf of $\mathcal{B}$,

$$\pi^M(g) := \pi(g) + g \circ M(x) \qquad (x \in X,\ g \in \mathcal{R}_x)$$

defines a sheaf homomorphism

$$\pi^M : \mathcal{R} \longrightarrow Q.$$

For every $U \in \mathfrak{U}^M$ and bounded $f \in \mathcal{R}_U$ we have

$$\pi^M r_U f = \rho_U \pi_U (f + K_U^M f):$$

Using $p \in \mathcal{P}_U^o$ with $M|U = \rho_U p$ we obtain for every $x \in U$

$$\pi^M r_U f(x) = \pi^M(r_U^x f) = \pi(r_U^x f) + (r_U^x f) \circ (\rho_U^x p)$$

$$= (\pi r_U f + (r_U f) \circ (\rho_U p))(x) = (\rho_U \pi_U f + \rho_U (f \circ p))(x)$$

$$= \rho_U (\pi_U f + K_U^M f)(x) = \rho_U \pi_U (f + K_U^M f)(x).$$

<u>PROPOSITION 6.8</u>: Let U be an open subset of X and $f \in \mathcal{C}_U$. Then $f \in {}^M\mathcal{S}_U$ if and only if $f \in \tilde{\mathcal{R}}_U$ and $\pi^M r_U f \in \Gamma^+(U,Q)$.

<u>PROOF</u>: Let $\mathfrak{W}$ be the set of all $V \in \mathfrak{U}^M$ such that $\overline{V} \subset U$ and $r_U^V f \in \mathcal{R}_V$. For every $V \in \mathfrak{W}$ we define

$$s_V = r_U^V (f + K_V^M f) = r_U^V f + K_V^M r_U^V f$$

and have

$$\pi^M r_V (r_U^V f) = \rho_V \pi_V s_V.$$

Suppose first that $f \in {}^M\mathcal{S}_U$. Then $f \in {}^M\tilde{\mathcal{R}}_U = \tilde{\mathcal{R}}_U$ by proposition 3.1 and corollary 6.6, and $\mathfrak{W}$ is a base of U. For every $V \in \mathfrak{W}$ we have $s_V \in \mathcal{S}_V \cap \mathcal{B}_V$ by the preceding lemma, hence $s_V \in \mathcal{H}_V + \mathcal{P}_V$, $\pi_V s_V \in \mathcal{P}_V$ and $\pi^M r_U f \mid V = \pi^M r_V(r_U^V f) = \rho_V \pi_V s_V \in \Gamma^+(V,Q)$. Therefore $\pi^M r_U f \in \Gamma^+(U,Q)$.

Assume now conversely that $f \in \breve{\mathcal{R}}_U$, $\pi^M r_U f \in \Gamma^+(U,Q)$ and consider $V \in \mathcal{W}$ such that there exists a $p \in \mathcal{P}_V$ with $\pi^M r_U f \mid V = \rho_V p$. Then $\pi_V s_V = p$, hence $s_V \in \mathcal{S}_V$. Since these sets V form a base of U, we obtain $f \in {}^M\mathcal{S}_U$ from lemma 6.7.

COROLLARY 6.9: The sheaf ${}^M\mathcal{H}$ of germs of M-harmonic functions is the kernel of the sheaf homomorphism $\pi^M : \mathcal{R} \longrightarrow Q$.

PROPOSITION 6.10: π^M is an epimorphism.

PROOF: Consider $U \in \mathfrak{U}^M$ and a bounded $q \in Q_U$. Defining

$$f := (I + K_U^M)^{-1} q$$

we have $f = q - K_U^M f \in Q_U \subset \mathcal{R}_U$ and

$$\pi^M r_U f = \rho_U \pi_U (f + K_U^M f) = \rho_U \pi_U q = \rho_U q.$$

This establishes the proposition.

Hence we have the following result which will be of importance in the next section.

COROLLARY 6.11: $0 \longrightarrow {}^M\mathcal{H} \longrightarrow \mathcal{R} \xrightarrow{\pi^M} Q \longrightarrow 0$ is a fine resolution of ${}^M\mathcal{H}$.

The rest of this section is mainly devoted to prove the existence of a section $M \in \Gamma^+(X,Q)$ such that $\{G_U^M\}_{U \in \mathfrak{U}^M}$ is strong-harmonic.

PROPOSITION 6.12: For every strictly positive $f \in \tilde{\mathcal{R}}_X$ there exists a section $M \in \Gamma^+(X,Q)$ such that f is M-superharmonic.

PROOF: Let $f \in \tilde{\mathcal{R}}_X$ be strictly positive. Since $\pi r_X f \in \Gamma(X,Q)$ there exist $M_1, M_2 \in \Gamma^+(X,Q)$ such that

$$\pi r_X f = M_1 - M_2$$

(proposition 4.9). Take

$$M: = \frac{1}{f} \circ M_2.$$

Then $M \in \Gamma^+(X,Q)$ and

$$\pi^M r_X f = \pi r_X f + f \circ M$$

$$= (M_1 - M_2) + M_2 = M_1 \in \Gamma^+(X,Q).$$

Therefore $f \in {}^M\mathcal{S}_X$ by proposition 6.8.

COROLLARY 6.13: $\tilde{\mathcal{R}}_X$ is the union of all ${}^M\mathcal{S}_X^+ - {}^M\mathcal{S}_X^+$ where $M \in \Gamma^+(X,Q)$.

PROOF: We know from proposition 3.1 and corollary 6.6 that for every $M \in \Gamma^+(X,Q)$

$${}^M\mathcal{S}_X^+ - {}^M\mathcal{S}_X^+ \subset {}^M\tilde{\mathcal{R}}_X = \tilde{\mathcal{R}}_X.$$

Consider now $f \in \tilde{\mathcal{R}}_X$. According to the remark following proposition 3.1 we have $f^+, f^- \in \tilde{\mathcal{R}}_X$. By lemma 3.4 (or the proof of corollary 3.5) there exists a strictly positive $g \in \tilde{\mathcal{R}}_X$. Then $g_1: = f^+ + g$ and $g_2: = f^- + g$ are strictly positive functions in $\tilde{\mathcal{R}}_X$. Hence there are $M_1, M_2 \in \Gamma^+(X,Q)$ such that $g_1 \in {}^{M_1}\mathcal{S}_X^+$, $g_2 \in {}^{M_2}\mathcal{S}_X^+$. Choosing $M = M_1 + M_2$ we obtain $g_1, g_2 \in {}^M\mathcal{S}_X^+$ (proposition 6.5) and hence $f = g_1 - g_2 \in {}^M\mathcal{S}_X^+ - {}^M\mathcal{S}_X^+$.

REMARK: In connection with theorem 5.3 the corollary 6.16 will show that for compact X we even have $\tilde{\mathcal{R}}_X = {}^M\mathcal{P}_X - {}^M\mathcal{P}_X$, if M is properly chosen.

LEMMA 6.14: There exist strictly positive global sections M in Q, i.e. $M \in \Gamma^+(X,Q)$ such that for all $U \in \mathfrak{U}^M$ the potential $p \in \mathcal{P}_U$ satisfying $\rho_U p = M|U$ is strictly positive on U.

PROOF: Let $\{U_i\}_{i \in I}$ be a locally finite covering of X consisting of sets in $\mathfrak{U}$ and let $\{V_i\}_{i \in I}$ be a shrinking of $\{U_i\}_{i \in I}$.

Let $i \in I$. Since U_i is strong-harmonic, there exists a strong potential $q_i \in \mathcal{P}_{U_i}$ ([5], 14.4). Take $\varphi_i \in \mathcal{C}_{U_i}^+$ such that

$\varphi_i \leq q_i$, $\varphi_i = q_i$ on V_i, $S(\varphi_i) \subset U_i$ and define $p_i = R_{\varphi_i}$. Then $p_i \in \mathcal{P}_{U_i}$, $C(p_i) \subset U_i$ and $H_V p_i(x) < p_i(x)$ for every $x \in V_i$ and $V \in \mathfrak{U}$ such that $x \in V$ and $\overline{V} \subset U_i$ ([5], 12.5). Thus $N_i := \rho_{U_i} p_i$ is a section in $\Gamma^+(U_i,Q)$ which is zero on $U_i \setminus C(p_i)$ and hence can be extended by zero to a section M_i in $\Gamma^+(X,Q)$. Choose

$$M = \sum_{i \in I} M_i \; .$$

Consider $U \in \mathfrak{U}^M$, $p \in \mathcal{P}_U$ with $\rho_U p = M|U$ and $x \in U$. Since $\{U_i\}_{i \in I}$ is locally finite, the set $I' = \{i \in I : U \cap U_i \neq \emptyset\}$ is finite. We have

$$M|U = \sum_{i \in I'} M_i \mid U$$

because every M_i has support in U_i. Since $\{V_i\}_{i \in I}$ is a covering of X, there exists an $i_o \in I$ with $x \in V_{i_o}$. Then $x \in U_{i_o}$ and hence $i_o \in I'$. Taking $V \in \bigcap_{i \in I'} \mathfrak{U}^{M_i}$ such that $x \in V$ and $\overline{V} \subset U \cap U_{i_o}$ there exist $p_i' \in \mathcal{P}_V$ with $\rho_V p_i' = M_i|V$ $(i \in I')$. Then

$$p \geq \rho_U^V p = \sum_{i \in I'} p_i' \geq p_{i_o}' .$$

But we have from our construction of p_{i_o} :

$$p_{i_o}'(x) = \rho_{U_{i_o}}^V p_{i_o}(x) = p_{i_o}(x) - H_V p_{i_o}(x) > 0 .$$

Hence $p(x) > 0$.

PROPOSITION 6.15: Let $M_o \in \Gamma^+(X,Q)$ such that there exists a strictly positive $s \in {}^{M_o}\mathcal{S}_X$. Then $\{G_U^{M_o+M_1}\}_{U \in \mathfrak{U}^{M_o+M_1}}$ is strong-harmonic for every strictly positive global section M_1 of Q.

PROOF: Let $M_1 \in \Gamma^+(X,Q)$ be strictly positive. Then lemma 6.1,3 implies that for every $U \in \mathfrak{U}^{M_o} \cap \mathfrak{U}^{M_o+M_1}$ and $x \in U$

$$G_U^{M_o+M_1} s(x) < G_U^{M_o} s(x) \leq s(x) .$$

So the statement follows from [5], 12.3.

REMARK: If there is a strictly positive $s \in \mathcal{S}_X^+$, then things like balayage of measures can be done using small perturbations: Take a strictly positive global section M in Q, work with $\{G_U^{\alpha M}\}$ for $\alpha > 0$ and let α tend to zero.

COROLLARY 6.16: There exists an $M \in \Gamma^+(X,Q)$ such that $\{G_U^M\}_{U \in \mathfrak{U}^M}$ is strong-harmonic.

PROOF: We take a strictly positive $f \in \tilde{\mathcal{R}}_X$. By proposition 6.12, there exists an $M_o \in \Gamma^+(X,Q)$ such that $f \in {}^{M_o}\mathcal{S}_X$. By lemma 6.14, there exists a strictly positive $M_1 \in \Gamma^+(X,Q)$. Choosing $M = M_o + M_1$ we conclude from proposition 6.15 that $\{G_U^M\}_{U \in \mathfrak{U}^M}$ is strong-harmonic.

7. The index-zero theorem

From now on we suppose that X is compact. We want to show that $\dim H^o(X,\mathcal{H}) = \dim H^1(X,\mathcal{H}) < \infty$. As we observed earlier this means that the mapping d from $\Gamma(X,\mathcal{R})$ into $\Gamma(X,Q)$ induced by π is a Fredholm operator of index zero.

For the following considerations we choose a section $M \in \Gamma^+(X,Q)$ such that $\{G_U^M\}_{U \in \mathfrak{U}^M}$ is strong-harmonic. We know by corollary 6.16 that this is possible. Let

$$d^M : \Gamma(X,\mathcal{R}) \longrightarrow \Gamma(X,Q)$$

be the mapping induced by $\pi^M : \mathcal{R} \longrightarrow Q$, i.e. $d^M(\sigma) = \pi^M\sigma$. Since $0 \longrightarrow {}^M\mathcal{H} \longrightarrow \mathcal{R} \xrightarrow{\pi^M} Q \longrightarrow 0$ is a fine resolution of ${}^M\mathcal{H}$ (corollary 6.11), we have on one hand

$$H^o(X, {}^M\mathcal{H}) \cong \ker d^M ,$$

$$H^1(X, {}^M\mathcal{H}) \cong \operatorname{coker} d^M$$

whereas on the other hand $H^o(X,{}^M\mathcal{H}) = H^1(X, {}^M\mathcal{H}) = 0$ by theorem 5.3. Thus d^M is an isomorphism from $\Gamma(X,\mathcal{R})$ onto $\Gamma(X,Q)$.

Recalling that $\tilde{r}_X$ is the natural isomorphism from the subspace $\tilde{\mathcal{R}}_X$ of $\mathcal{C}_X$ onto $\Gamma(X,\mathcal{R})$ we see that $d^M\tilde{r}_X$ is an isomorphism from $\tilde{\mathcal{R}}_X$ onto $\Gamma(X,Q)$. Defining

$$Tf := (d^M\tilde{r}_X)^{-1}\,(f\circ M) \qquad (f \in \mathcal{B}_X)$$

we therefore obtain an operator T on $\mathcal{B}_X$ with $T(\mathcal{B}_X) \subset \tilde{\mathcal{R}}_X$. By definition of π^M we have for every $\sigma \in \Gamma(X,\mathcal{R})$

$$\begin{aligned} d(\sigma) &= d^M(\sigma) - \sigma\circ M = d^M(\sigma) - (\tilde{r}_X^{-1}(\sigma))\circ M \\ &= d^M\tilde{r}_X(I-T)\tilde{r}_X^{-1}(\sigma) , \end{aligned}$$

i.e.

$$d = (d^M\tilde{r}_X)\,(I-T)\tilde{r}_X^{-1} .$$

Since $\tilde{r}_X^{-1}$ is an isomorphism from $\Gamma(X,\mathcal{R})$ onto $\tilde{\mathcal{R}}_X$ and $d^M\tilde{r}_X$ an isomorphism from $\tilde{\mathcal{R}}_X$ onto $\Gamma(X,Q)$, we hence have to prove that $(I-T)\,|\tilde{\mathcal{R}}_X$ is a Fredholm operator of index zero on $\tilde{\mathcal{R}}_X$. To that end we show the following

<u>PROPOSITION 7.1</u>: T is a strong Feller kernel on X.

<u>PROOF</u>: Since $T(\mathcal{B}_X) \subset \tilde{\mathcal{R}}_X \subset \mathcal{C}_X$ we have only to show that T is a kernel.

1. We have $T(\mathcal{B}_X^+) \subset {}^M\mathcal{S}_X$, in particular T is positive: For every $f \in \mathcal{B}_X^+$

$$\pi^M r_X(Tf) = (d^M\tilde{r}_X)\,(Tf) = f\circ M \in \Gamma^+(X,Q),$$

hence $Tf \in {}^M\mathcal{S}_X$ by proposition 6.8. But since X is compact and $\{G_U^M\}_{U\in\mathfrak{U}^M}$ is strong-harmonic, every M-superharmonic function on X is non-negative by [5], 13.2.

2. T is σ-continuous: Let (f_n) be a decreasing sequence in $\mathcal{B}_X^+$ with $\inf f_n = 0$. Define

$$t_n = Tf_n, \qquad t = \inf t_n$$

and

$$s_n = \sum_{k=n}^{\infty} T(f_k - f_{k+1}) = t_n - t.$$

By (1), (t_n) is a decreasing sequence in ${}^M\mathcal{S}_X^+$, hence t is non-negative and nearly M-hyperharmonic and (s_n) is a decreasing sequence in ${}^{M*}\mathcal{H}_X^+$. The equation $t + s_n = t_n$ therefore implies that $t \in {}^M\mathcal{S}_X$ and $s_n \in {}^M\mathcal{S}_X$ for every n. So we obtain from proposition 6.8 that $t, s_n \in \tilde{\mathcal{R}}_X$ and $\pi^M r_X t$, $\pi^M r_X s_n \in \Gamma^+(X,Q)$ where

$$\pi^M r_X t + \pi^M r_X s_n = \pi^M r_X t_n = f_n \circ M.$$

Hence $\pi^M r_X t = 0$ by proposition 4.11, i.e. $d^M\tilde{r}_X(t) = 0$ and therefore $t = 0$.

Giving $\mathcal{B}_X$ the supremum norm we have the following

<u>COROLLARY 7.2</u>: T^2 is a compact operator on $\mathcal{B}_X$ and hence $I - T$ a Fredholm operator of index zero.

<u>PROOF</u>: An elementary proof for the fact that the composed of two strong Feller kernels on X is compact is given in [10]. p.172. So T^2 is compact by the preceding proposition. For every $\alpha \geq 0$ we have

$$(I-\alpha T)(I+\alpha T) = (I+\alpha T)(I-\alpha T) = I - \alpha^2 T^2$$

where $\alpha^2 T^2$ is a compact operator and hence $I - \alpha T$ is a Fredholm operator. Since I has index zero and the index is a continuous function we conclude that $I - T$ has index zero.

<u>THEOREM 7.3</u>: d is a Fredholm operator of index zero, i.e. $\dim H^0(X,\mathcal{H}) = \dim H^1(X,\mathcal{H}) < \infty$.

<u>PROOF</u>: We already reduced the statement to the question whether $(I-T)|\tilde{\mathcal{R}}_X$ is a Fredholm operator of index zero on $\tilde{\mathcal{R}}_X$. But $T(\mathcal{B}_X) \subset \tilde{\mathcal{R}}_X$ implies that $\ker (I-T)|\tilde{\mathcal{R}}_X = \ker (I-T)$ and $\tilde{\mathcal{R}}_X/(I-T)(\tilde{\mathcal{R}}_X)$ is isomorphic to $\mathcal{B}_X/(I-T)(\mathcal{B}_X)$ by $f + (I-T)(\tilde{\mathcal{R}}_X) \longrightarrow f + (I-T)(\mathcal{B}_X)$. So corollary 7.2 establishes the theorem.

Bibliography

H.BAUER [1] : Harmonische Räume und ihre Potentialtheorie. Lecture Notes in Mathematics 22, Berlin-Heidelberg-New York: Springer 1966.

J.-M.BONY [2] : Opérateurs elliptiques dégénérés. Ann.Inst.Fourier 19/1 (1969), 277-304.

G.E.BREDON [3] : Sheaf theory. McGraw-Hill 1967.

R.GODEMENT [4] : Topologie algébrique et théorie des faisceaux. Paris: Hermann 1958.

W.HANSEN [5] : Potentialtheorie harmonischer Kerne. Seminar über Potentialtheorie, Lecture Notes in Mathematics 69, 103-159, Berlin-Heidelberg-New York: Springer 1968.

L.L.HELMS [6] : Introduction to potential theory. Wiley-Interscience 1969.

R.-M.HERVE [7] : Recherches axiomatiques sur la théorie des fonctions surharmoniques et du potentiel. Ann.Inst.Fourier 12 (1962), 415-571.

R.-M.HERVE M.HERVE [8] : Les fonctions surharmoniques associées à un opérateur elliptique du second ordre à coefficients discontinus. Ann.Inst.Fourier 19/1 (1968), 305-359.

R.KULTZE [9] : Garbentheorie. Stuttgart: Teubner 1970.

P.-A.MEYER [10] : Séminaire de probabilités II. Lecture Notes in Mathematics 51, Berlin-Heidelberg-New York: Springer 1968.

B.WALSH [11] : Flux in axiomatic potential theory. I. Cohomology. Inventiones math. 8 (1969), 175-221.

[12] Flux in axiomatic potential theory. II. Duality. Ann.Inst.Fourier 19/2, (1969), 371-417.

[13] Perturbation of harmonic structures and an index-zero theorem. Ann.Inst.Fourier 20,1 (1970), 317-359.

B.WALSH [14] : Operator theory of degenerate elliptic-parabolic equations.
International Symposium on Operator Theory, Indiana 1970.

Contents

MARTIN BOUNDARY AND $\mathscr{H}^p$-THEORY OF HARMONIC SPACES

by

Klaus Janßen

0. Introduction

Many results concerning holomorphic or complex-valued harmonic functions in the unit disc of the plane are special cases of theorems in a more general set up. For example, Mme. L.LUMER-NAIM [16] developed for Brelot harmonic spaces an $\mathscr{H}^p$-theory which generalizes the classical theory for the unit disc (e.g. K.HOFMANN [13]).

Among others she obtained theorems of Phragmen-Lindelöf type (in which it is proved that certain subharmonic functions are bounded or increase 'very rapidly' approaching a suitable boundary) and an F. and M.Riesz theorem (integral representation of a class of harmonic functions by means of measures which are absolutely continuous with respect to a given measure).

The aim of this paper is to transfer results of Mme. L.LUMER-NAIM to harmonic spaces which satisfy the axioms of H.BAUER [1] and so to give a positive answer to a question raised in [16]. Therefore, it is necessary to generalize in an adequate manner the fundamental concepts of 'Martin boundary' and 'uniform integrability'. This will be done by the introduction of reference measures (cf. definition (1.1)) which replace the Dirac measures in the case of a Brelot harmonic space. An essential property of a reference measure r is the validity of the following convergence theorem:

The upper envelope of an increasing sequence of

r-integrable harmonic functions is harmonic if it is r-integrable.

In general, the organization of this paper follows Mme. L.LUMER-NAIM [16]. After some preliminaries, the Martin boundary of a harmonic space is introduced to be the set of the extreme rays of the cone of positive harmonic functions, equipped with a canonical topology. For a given reference measure r, we develop an integral representation for the positive r-integrable harmonic functions by means of regular Borel measures ('representing measures') on the Martin boundary. Following M.SIEVEKING [21], we consider a Dirichlet problem relative to r for functions defined on the Martin boundary.

Later on the following generalization of a theorem of DOOB (cf. [7]) is proved:

Assume that the constant functions are harmonic. A positive harmonic function f is 'r-uniformly integrable' if and only if f is the Dirichlet solution relative to r for some positive function on the Martin boundary.

In the following two chapters, we consider sets $\mathcal{H}^{\Phi}(r)$ of harmonic functions. For some positive function Φ on $\mathbb{R}_+$, $\mathcal{H}^{\Phi}(r)$ denotes the set of those harmonic functions f, for which $\Phi(|f|)$ is subharmonic and dominated by an r-integrable harmonic function. Especially, in the case $\Phi = \Phi_p : t \longrightarrow t^p (p > 1)$, $\mathcal{H}^{\Phi_p}(r)$ is isometric to the Banach space $L^p(\tilde{\mu}_1)$, where $\tilde{\mu}_1$ is the representing measure of the constant function 1. Analogous statements hold for a wide class of convex functions Φ.

In the final chapter we obtain (as in [16], IV) an F. and M.Riesz theorem and a Phragmen-Lindelöf principle which is compared with a result of M.H.PROTTER, H.F.WEINBERGER [18] for the heat equation.

I would like to express my gratitude to B.Anger,

J.Bliedtner, J.Lembcke in Erlangen and M.Sieveking in Zürich for many interesting talks. Especially, I am indebted to Professor H.Bauer for his comprehensive aid.

1. Preliminaries

We always assume that $(X,\mathcal{H})$ is a harmonic space in the sense of H.BAUER [1], i.e. X is a locally compact space with a countable base of open sets and $\mathcal{H}$ is a sheaf of vector spaces of real-valued continuous functions satisfying the axioms I - IV of [1]. We shall use the same notations as in [1].

First of all we shall prove some additional properties of $(X,\mathcal{H})$.

1.1 DEFINITION: A positive Radon measure r on X is called reference measure (relative to $(X,\mathcal{H})$) iff X is the smallest absorbing set containing the support of r. Moreover, if $r(1) = 1$, we call r a normed reference measure.

For a given harmonic function h, we can always construct a normed reference measure r, such that h is r-integrable: We only have to sum up suitable point-measures of points of a countable dense subset of X.

In a connected Brelot space $(X,\mathcal{H})$, there exist only trivial absorbing sets, hence any non-trivial positive Radon measure on X (especially the Dirac measure of a point in X) is a reference measure relative to $(X,\mathcal{H})$.

1.2 LEMMA: Let r be a reference measure relative to $(X,\mathcal{H})$. A sequence (h_n) of positive harmonic functions converges locally uniformly to zero provided $\lim_{n\to\infty} \int h_n \, dr = 0$.

Especially, a positive harmonic function h on X equals

zero iff $\int h\,dr = 0$.

PROOF: The first statement is a consequence of Harnack's inequality: For every compact subset K of X, there exists a real number $a_K(r)$ such that

$$\sup h(K) \leq a_K(r) \int h\,dr$$

for every positive harmonic function h on X ([1], 1.1.4). Applying this result to the constant sequence (h) we obtain the second assertion. ⌟

1.3 DEFINITION: A sequence (U_n) of open relatively compact subsets of X is called an exhaustion of X iff (U_n) increases to X such that $\overline{U}_n \subset U_{n+1}$ for every $n \in \mathbb{N}$.

1.4 LEMMA: Let (U_n) be an exhaustion of X and let r be a reference measure. If h and h_n $(n \in \mathbb{N})$ are r-integrable numerical functions on X such that

i) h_n is harmonic on U_n $(n \in \mathbb{N})$,

ii) $(\mathrm{rest}_{U_n} h_{n+i})_{i \in \mathbb{N}}$ is an upward filtering sequence converging pointwise to h on U_n $(n \in \mathbb{N})$,

then h is harmonic on X.

PROOF: Obviously it is sufficient to prove the harmonicity of the restriction of h to U_n for an arbitrarily chosen $n \in \mathbb{N}$.

As a limit of an upward filtering sequence of harmonic functions, $\mathrm{rest}_{U_n} h$ is hyperharmonic on U_n. Hence the convergence axiom gives the assertion if $D := \{x \in U_n : h(x) < \infty\}$ is a dense subset of U_n. But, since $\int h\,dr < \infty$, $A_h := \overline{\{x \in X : h(x) < \infty\}}$ is an absorbing set containing the support of r, i.e. $A_h = X$ and consequently D is dense in U_n. ⌟

1.5. LEMMA: Assume (U_n) is an exhaustion, r is a reference measure, and s is an r-integrable subharmonic function on X. Define for any $n \in \mathbb{N}$ a function $s_n : X \longrightarrow \overline{\mathbb{R}}$ by

$$s_n(x) := \begin{cases} \int s \, d\mu_x^{U_{n+1}} & \text{if } x \in \overline{U}_n \\ s(x) & \text{if } x \notin \overline{U}_n \end{cases} .$$

Then (s_n) is an increasing sequence, and $u := \sup s_n$ is r-integrable if and only if there is a real number M such that

$$\int_{\overline{U}_n} s_n \, dr \leq M \quad \text{for every } n \in \mathbb{N}.$$

In this case, $\int u \, dr = \lim \int_{\overline{U}_n} s_n \, dr$ and u is the smallest harmonic majorant of s.

PROOF: For every $n \in \mathbb{N}$ the function

$$x \longrightarrow \int s \, d\mu_x^{U_n} \qquad (x \in U_n)$$

ist the smallest harmonic majorant of $\mathrm{rest}_{U_n} s$, whence (s_n) is an increasing sequence such that $u = \sup s_n \geq s$.

Since $\mathrm{rest}_{U_n} s_n$ is harmonic on U_n $(n \in \mathbb{N})$, application of (1.4) yields: $u = \sup s_n$ is harmonic if u is r-integrable. In this case u is the smallest harmonic majorant of s, since for every harmonic majorant h of s

$$\int s \, d\mu_x^{U_{n+1}} \leq \int h \, d\mu_x^{U_{n+1}} = h(x) \qquad (n \in \mathbb{N},\ x \in \overline{U}_n),$$

especially

$$s_n(x) \leq h(x) \qquad (n \in \mathbb{N},\ x \in \overline{U}_n)$$

and finally

$$u = \sup s_n \leq h.$$

Furthermore, if u is r-integrable, we obtain for every $n \in \mathbb{N}$

$$\int_{\overline{U}_n} s_n \, dr \leq \int_{\overline{U}_n} u \, dr \leq \int_{\overline{U}_n} |u| dr \leq \int |u| dr < \infty .$$

Conversely, assume

$$\int_{\overline{U}_n} s_n \, dr \leq M \quad \text{for every } n \in \mathbb{N}.$$

Then the r-integrability of s and the relation

$$\sup_n 1_{\overline{U}_n} \cdot (s_n - s) = u - s$$

imply the r-integrability of u by B.LEVI's convergence theorem and furthermore

$$\lim_{n\to\infty} \int_{\overline{U}_n} s_n \, dr = \int u \, dr .$$

The proofs of the following two lemmata are exactly the same as the proofs of [16], III. lemma 9, II. lemma 3.

1.6 LEMMA: Let u be subharmonic in X and let Φ be an increasing convex real-valued function on an intervall $I \subset \mathbb{R}$ such that $u(X) \cup \{0\} \subset I$. Then $\Phi(u)$ is subharmonic in each of the following two cases:

i) the constant function 1 is harmonic;

ii) the constant function 1 is superharmonic and $\Phi(0) \leq 0$.

1.7.LEMMA: Assume the constant function 1 is superharmonic and let $u_1,\ldots,u_n$ be harmonic (or positive subharmonic) functions on X. Then, for any real number $p \geq 1$, the function $(\sum_{i=1}^{n} u_i^2)^{p/2}$ is subharmonic on X.

2. Martin boundary

In this chapter we develop an integral representation of those positive harmonic functions which are r-integrable for a given reference measure r. By and large, we follow M.SIEVEKING [21], except that we use a more general concept of a reference measure. Especially, in our set up, a Dirac measure is a reference measure for a connected Brelot space. Later, we consider a Dirichlet problem for the Martin boundary. According to [21], we obtain results, received for Brelot spaces by K.GOWRISANKARAN ([11], [12]).

Let $(X,\mathcal{H})$ be a harmonic space in the sense of H.BAUER [1] and let ${}_+\mathcal{H}_X$ be the set of positive harmonic functions on X.

Endowed with the topology of uniform convergence on compact subsets of X, ${}_+\mathcal{H}_X$ is a complete metrizable space with a countable base of open sets.

2.1 DEFINITION: For any reference measure r, we define the following sets:

$$\mathcal{H}_r^1 := \{h \in {}_+\mathcal{H}_X : \int h\, dr \leq 1\} ,$$

$$\Delta^r := \{h \in {}_+\mathcal{H}_X : \int h\, dr = 1\} ,$$

$$\Delta_e := \{h \in {}_+\mathcal{H}_X : h \neq 0 \text{ lies on an extreme ray of } {}_+\mathcal{H}_X\} ,$$

$$\Delta_e^r := \Delta^r \cap \Delta_e .$$

2.2 REMARK: For connected Brelot spaces, the Dirac measure of a point of X is a reference measure. In this case, Δ^r is a compact base of the cone ${}_+\mathcal{H}_X$.

In general, this is no longer true as the following example shows:

2.3 EXAMPLE: Let $\mathcal{H}$ be the sheaf of solutions u of the heat equation $\frac{\partial^2 u}{\partial x^2} = \frac{\partial u}{\partial t}$ in $X := \mathbb{R} \times]0,\infty[$. $(X,\mathcal{H})$ is a harmonic space. Define

$\lambda :=$ restriction of the Lebesgue measure on $\mathbb{R}$ to $\mathbb{R}_+$,

$$f(t) := \begin{cases} 0 & \text{if } 0 < t \leq \frac{1}{4} \\ \frac{1}{4}\, t^{-3/2} & \text{if } \frac{1}{4} < t < \infty \end{cases} ,$$

and

$$r := \varepsilon_o \otimes (f\lambda).$$

The support of the measure r on X is the set $\{0\} \times [\frac{1}{4}, \infty[$. Since every absorbing set in X is given by X or $\mathbb{R} \times]0,t_o]$ with $t_o \in \mathbb{R}_+$, r is a reference measure relative to $(X,\mathcal{H})$. Since

$$\int 1\, dr = \frac{1}{4} \int_{\frac{1}{4}}^{\infty} t^{-3/2}\, dt = \frac{1}{4} [-2t^{-1/2}]_{\frac{1}{4}}^{\infty} = 1 ,$$

r is a normed reference measure.

According to M.SIEVEKING ([21], p.59), every function on an extreme ray of ${}_+\mathcal{H}_X$ is given by

$$(x,t) \longrightarrow ct^{-1/2} \cdot \exp\left(-\frac{(x-a)^2}{4t}\right) \qquad ((x,t) \in X,\ a \in \mathbb{R},\ c \in \mathbb{R}_+).$$

For every $a \in \mathbb{R}$ define the function $h_a : X \longrightarrow \mathbb{R}$ by

$$h_a(x,t) := \begin{cases} \dfrac{a^2}{1-\exp(-a^2)} \cdot t^{-1/2} \cdot \exp\left(-\dfrac{(x-a)^2}{4t}\right), & \text{if } a \neq 0 \\ t^{-1/2} \cdot \exp\left(-\dfrac{x^2}{4t}\right), & \text{if } a = 0 \end{cases}.$$

For $a \neq 0$, we obtain

$$\int h_a \, dr = \frac{1}{4} \cdot \frac{a^2}{1-\exp(-a^2)} \cdot \int_{\frac{1}{4}}^{\infty} t^{-2} \cdot \exp\left(-\frac{a^2}{4t}\right) dt$$

$$= \frac{1}{4} \cdot \frac{a^2}{1-\exp(-a^2)} \cdot \left[\frac{4}{a^2} \cdot \exp\left(-\frac{a^2}{4t}\right) \right]_{\frac{1}{4}}^{\infty} = 1 ;$$

moreover, $\int h_o \, dr = \frac{1}{4} \cdot \int_{\frac{1}{4}}^{\infty} t^{-2} dt = 1$, i.e. every harmonic function which lies on an extreme ray of ${}_+\mathcal{H}_X$ is r-integrable and we have

$$\Delta_e^r = \{h_a : a \in \mathbb{R}\}.$$

Let (a_n) be a sequence in $\mathbb{R}$ tending to infinity. Then $(h_{a_n}) \subset \Delta_e^r \subset \Delta^r$ and $h_{a_n} \longrightarrow 0$ uniformly on compact subsets of X. Since $0 \notin \Delta^r$, Δ^r is not closed in ${}_+\mathcal{H}_X$.

<u>2.4 PROPOSITION</u>: For every reference measure r, the set $\mathcal{H}_r^1$ is a simplex and a cap of ${}_+\mathcal{H}_X$.

<u>PROOF</u>: a) First of all, we prove the compactness of $\mathcal{H}_r^1$. For each $x \in X$, Harnack's inequality yields the existence of a constant $\alpha_x(r)$ depending only on x and r such that

$$0 \leq h(x) \leq \alpha_x(r) \cdot \int h \, dr \leq \alpha_x(r)$$

for every $h \in \mathcal{H}_r^1$, i.e. $\{h(x) : h \in \mathcal{H}_r^1\}$ is a bounded subset of $\mathbb{R}$. Moreover, $\mathcal{H}_r^1$ is equicontinuous on X (cf. H.BAUER [1], 4.6.3). Hence we obtain by Ascoli's theorem that $\mathcal{H}_r^1$ is relatively compact in ${}_+\mathcal{H}_X$.

Now, if (h_n) is a sequence in $\mathcal{H}_r^1$ converging to $h \in {}_+\mathcal{H}_X$, then Fatou's lemma implies

$$\int h\, dr \leq \liminf \int h_n\, dr \leq 1 \ ,$$

i.e. $h \in \mathcal{H}_r^1$, whence $\mathcal{H}_r^1$ is closed in ${}_+\mathcal{H}_X$ and consequently compact.

b) The map $\quad h \longrightarrow \int h\, dr$

from ${}_+\mathcal{H}_X$ into $\overline{\mathbb{R}}_+$ is additive, positive homogeneous and lower semicontinuous (Fatou's lemma). According to R.R.PHELPS ([19], prop. 11.2) $\mathcal{H}_r^1$ is a cap of ${}_+\mathcal{H}_X$.

Moreover, ${}_+\mathcal{H}_X$ is a lattice in its own order, hence $\mathcal{H}_r^1$ is a simplex ([19], prop. 11.3). $\lrcorner$

Let E be a topological space. We denote by $\underline{\mathcal{M}_+(E)}$ the set of positive finite regular Borel measures on E, where a <u>regular Borel measure</u> on E is defined as a measure on the σ-algebra of the Borel sets in E which is finite on the compact sets and inner regular with respect to the system of compact subsets of E. $\mu \in \mathcal{M}_+(E)$ is said to be <u>supported by a subset $A \subset E$</u> iff every compact set $K \subset E$ disjoint from A is a μ-nullset.

<u>2.5 THEOREM</u>: Let r be a reference measure on X. For every function $h \in \Delta^r$, there is a unique measure $\mu_h \in \mathcal{M}_+(\Delta^r)$ having the properties

i) $\mu_h(\Delta^r) = 1$, μ_h is supported by the G_δ-set Δ_e^r;

ii) $L(h) = \int L(k)\, d\mu_h(k)$ for every semicontinuous affine function L on $\mathcal{H}_r^1$.

Conversly, let $\mu \in \mathcal{M}_+(\Delta^r)$ such that $\mu(\Delta^r) = \mu(\Delta_e^r) = 1$. Then a harmonic function $h \in \Delta^r$ is defined by

$$h(x) = \int k(x)\, d\mu(k) \qquad (x \in X).$$

<u>PROOF</u>: According to N.BOURBAKI ([6], p.62, prop.22), the gauge functional of $\mathcal{H}_r^1$ is given by

$$p_{\mathcal{H}_r^1}(f) := \inf\{a > 0 : f \in a \cdot \mathcal{H}_r^1\}$$
$$= \int f \, dr \qquad (f \in {}_+\mathcal{H}_X) .$$

Hence $\{0\} \cup \{\Delta_e^r\}$ is exactly the set of extreme points of $\mathcal{H}_r^1$ (cf. [6], p.111, cor. 1). Since the linear maps $L_x : {}_+\mathcal{H}_X \longrightarrow \mathbb{R}_+$, defined by $L_x(k) := k(x)$, are continuous, the assertions follow according to (2.4) from the existence and uniqueness theorem of Choquet-Meyer (cf. H.BAUER [2], 3.2.3, 3.2.4, 4.3.5, 2.5.2). _/

2.6 DEFINITION: On ${}_+\mathcal{H}_X \setminus \{0\}$ we define a binary relation $\underline{R}$ by:

$h_1 R h_2$ iff there is an $\alpha > 0$ such that $h_1 = \alpha h_2$
$(h_1, h_2 \in {}_+\mathcal{H}_X \setminus \{0\})$.

Since for every open set $U \subset {}_+\mathcal{H}_X \setminus \{0\}$, the saturated relative to R is given by the open set $\bigcup_{\lambda>0} \lambda U$, R is an open equivalence relation (cf. N.BOURBAKI [3], p.55, 5.2). In the sequel let

$$\underline{\mathcal{Q}} := {}_+\mathcal{H}_X \setminus \{0\} \,/\, R$$

be endowed with the quotient topology. Denote by $\underline{i}$ the canonical surjection from ${}_+\mathcal{H}_X \setminus \{0\}$ onto $\mathcal{Q}$ induced by R.

2.7 PROPOSITION: $\mathcal{Q}$ is a topological Hausdorff space.

PROOF: Let G(R) be the graph of the open equivalence relation R. According to N.BOURBAKI ([3], p.88, prop.8), we have to prove that G(R) is closed in $({}_+\mathcal{H}_X \setminus \{0\}) \times ({}_+\mathcal{H}_X \setminus \{0\})$. Since ${}_+\mathcal{H}_X$ is metrizable, we have to prove that for any sequence $(H_n) \subset G(R)$, converging to an element $H \in ({}_+\mathcal{H}_X \setminus \{0\}) \times ({}_+\mathcal{H}_X \setminus \{0\})$, we have $H \in G(R)$.

Indeed, let

$$H_n = (\lambda_n h_n, h_n) \qquad (n \in \mathbb{N},\ h_n \in {}_+\mathcal{H}_X \setminus \{0\},\ \lambda_n \in \mathbb{R}_+^*),$$
$$H = (g, h) \qquad (g, h \in {}_+\mathcal{H}_X \setminus \{0\})$$

such that (H_n) converges to H. Then the sequences $(\lambda_n h_n)$, (h_n)

converge uniformly on compact subsets of X to g, h respectively. Especially

$$\lim_{n\to\infty} \lambda_n h_n(x) = g(x)$$

and

$$\lim_{n\to\infty} h_n(x) = h(x)$$

for every $x \in X$. Hence, (λ_n) is convergent to some $\lambda \in \mathbb{R}_+$, since $h(x) \neq 0$ for some $x \in X$. Consequently,

$$\lambda = \frac{g(x)}{h(x)} \quad \text{for every } x \in X \text{ such that } h(x) \neq 0.$$

Then

$$\lambda \cdot h(x) = g(x) \quad \text{for every } x \in X,$$

since $h(x) = 0$ implies $g(x) = (\lim \lambda_n)\cdot(\lim h_n(x)) = 0$. This is the desired relation $g\, R\, h$, i.e. $H = (g, h) \in G(R)$. $\_/$

2.8 NOTATIONS and CONSEQUENCES: Let r be a reference measure on X.

a) Consider on the subsets $\tilde{\Delta}_e := i(\Delta_e)$ and $\tilde{\Delta}_e^r := i(\Delta_e^r)$ of $\mathcal{Q}$ the relative topology of $\mathcal{Q}$. $\tilde{\Delta}_e$ is called the Martin boundary of $(X,\mathcal{H})$.

b) Δ_e^r is a G_δ-set in $\mathcal{H}_r^1$, thus especially a Polish space. If we denote by i_r the restriction of i to Δ_e^r, then i_r is a continuous injection into $\tilde{\Delta}_e$. In many cases i_r is a homeomorphism from Δ_e^r onto $\tilde{\Delta}_e^r$; for example this is true if Δ^r is compact (this is true, if $(X,\mathcal{H})$ is a connected Brelot space and $r = \epsilon_x$ for some $x \in X$). Even in the example of the heat equation (2.25) we have a homeomorphism. But in general, it seems that this property will not hold.

c) In order to obtain a Martin boundary version of theorem (2.5), we have to consider images of measures:

For a given measure $\mu \in \mathcal{M}_+(\Delta_e^r)$, the image $i_r(\mu)$ is a positive finite regular Borel measure on the Martin boundary; $i_r(\mu)$ is supported by $\tilde{\Delta}_e^r$. Conversely, every $\tilde{\mu} \in \mathcal{M}_+(\tilde{\Delta}_e)$ which is supported by $\tilde{\Delta}_e^r$, is according to J.LEMBCKE [14] the image of some

$\mu \in \Delta_e^r$, i.e.

$$(*) \qquad \tilde{\mu} = i_r(\mu) .$$

For measures μ and $\tilde{\mu}$, satisfying (*), we have the following

2.9 PROPOSITION: Let f, $\tilde{f}$ be positive numerical functions defined on Δ_e^r, $\tilde{\Delta}_e$ respectively, such that $f(k) = \tilde{f}(i(k))$ for every $k \in \Delta_e^r$. Then f is μ-integrable if and only if $\tilde{f}$ is $\tilde{\mu}$-integrable, in which case $\int f\, d\mu = \int \tilde{f}\, d\tilde{\mu}$.

2.10 DEFINITION: For a reference measure r on X, define $\tilde{K}_x^r: \tilde{\Delta}_e \longrightarrow \mathbb{R}$ by

$$\tilde{K}_x^r(\tilde{k}) := \begin{cases} k(x) & \text{if } k \in \Delta_e^r,\ i(k) = \tilde{k} \\ 0 & \text{otherwise} \end{cases} .$$

Obviously, $\tilde{K}_x^r(i(k)) = k(x)$ for every $k \in \Delta_e^r$. Harnack's inequality implies that $\tilde{K}_x^r$ is bounded for every $x \in X$.

2.11 THEOREM: a) For every positive harmonic function h and for every reference measure r on X such that $\int h\, dr < \infty$, there exists a measure $\tilde{\mu}_h \in \mathcal{M}_+(\tilde{\Delta}_e)$ such that

i) $\tilde{\mu}_h$ is supported by $\tilde{\Delta}_e^r$,

ii) $h(x) = \int \tilde{K}_x^r\, d\tilde{\mu}_h$ for every $x \in X$.

If $\tilde{\nu} \in \mathcal{M}_+(\tilde{\Delta}_e)$ is another measure having properties i) and ii), then $\tilde{\mu}_h = \tilde{\nu}$.

b) Conversely, for every $\tilde{\mu} \in \mathcal{M}_+(\tilde{\Delta}_e)$, there is a reference measure r on X such that $\tilde{\mu}$ is supported by $\tilde{\Delta}_e^r$. Then

$$h(x) := \int \tilde{K}_x^r\, d\tilde{\mu} \qquad (x \in X)$$

defines a positive r-integrable harmonic function on X.

PROOF: It only remains to show that for every $\tilde{\mu} \in \mathcal{M}_+(\tilde{\Delta}_e)$ there exists a reference measure r on X such that $\tilde{\mu}$ is supported by $\tilde{\Delta}_e^r$, since according to (2.8) - (2.10) the other assertions of the theorem follow immediately by theorem (2.5).

Now suppose $\tilde{\mu} \in \mathcal{M}_+(\tilde{\Delta}_e)$. Since $\tilde{\mu}$ is regular, $\tilde{\mu}$ is supported by a set $\tilde{K}$ which is a countable union of compact subsets of $\tilde{\Delta}_e$. According to N.BOURBAKI ([4], p.45, prop. 18), there exists a sequence (K_n) of compact subsets of $_+\mathcal{H}_X \setminus \{0\}$ such that $i(\bigcup_{n=1}^{\infty} K_n) = \tilde{K}$. Now the following lemma (2.12) implies the existence of a reference measure r on X such that $\int h\, dr < \infty$ for every $h \in \bigcup_{n=1}^{\infty} K_n$. This implies $\tilde{K} \subset \tilde{\Delta}_e^r$: Indeed, for every $\tilde{k} \in \tilde{K}$, there is some $k \in \bigcup_{n=1}^{\infty} K_n$ satisfying $i(k) = \tilde{k}$. Then $\alpha k \in \Delta^r$ for a suitable $\alpha > 0$, hence $\tilde{k} = i(k) = i(\alpha k) \in \tilde{\Delta}^r$. It follows $\tilde{k} \in \tilde{\Delta}_e^r$, since $\tilde{K} \subset \tilde{\Delta}_e$. ⌟

<u>2.12 LEMMA</u>: For every sequence (K_n) of compact subsets of $_+\mathcal{H}_X$, there is a finite reference measure r on X such that

$$\int h\, dr < \infty \quad \text{for every} \quad h \in \bigcup_{n=1}^{\infty} K_n .$$

<u>PROOF</u>: Without loss of generality, assume the sequence (K_n) to be increasing. For any $x \in X$ and $n \in \mathbb{N}$ define $C_{x,n} := \{h(x) : h \in K_n\}$; $C_{x,n}$ is the continuous image of a compact set, hence a compact subset of $\mathbb{R}$.

Now let (x_n) be a countable dense subset of X and define for $n \in \mathbb{N}$

$$\alpha_n' := \sup \{h(x_n) : h \in K_n\},$$

$$\alpha_n := \sup \{\alpha_n', 1\} .$$

Then the measure $r := \sum_{n=1}^{\infty} 2^{-n} \alpha_n^{-1} \varepsilon_{x_n}$ satisfies all requirements: Obviously, r is a finite reference measure. If $h \in \bigcup_{n=1}^{\infty} K_n$, then $h \in K_{n_o}$ for some $n_o \in \mathbb{N}$, therefore

$$h(x_n) \leq \alpha_n \qquad \text{for every} \quad n \geq n_o$$

and consequently

$$\int h\, dr \leq \sum_{n=1}^{n_o-1} 2^{-n} \alpha_n^{-1} h(x_n) + \sum_{n=n_o}^{\infty} 2^{-n} < \infty .$$ ⌟

<u>2.13 DEFINITION</u>: Let r be a reference measure and let h be a positive r-integrable harmonic function on X. We call the unique measure $\tilde{\mu}_h$ on the Martin boundary, which exists according

to (2.11) the representing measure of h (relative to r).

In general, the representing measures of a harmonic function relative to different reference measures are not the same. However, there are some connections as the following remark shows:

2.14 REMARK: Let r_1, r_2 be reference measures on X. The map $f: \mathcal{H}_X^+ \longrightarrow \overline{\mathbb{R}}_+$ defined by $f(k) := \int k \, dr_2$ is lower semi-continuous (Fatou's lemma). Assume that f is finite and bounded on compact subsets of $\Delta_e^{r_1}$ (in the case of a connected Brelot space $(X,\mathcal{H})$, f is continuous if $r_2 = \varepsilon_x$ for some $x \in X$). Then $\tilde{\Delta}_e^{r_1} \subset \tilde{\Delta}_e^{r_2}$. Define

$$\tilde{f}(\tilde{k}) := \begin{cases} \dfrac{\int k \, dr_2}{\int k \, dr_1} & \text{if } k \in \tilde{k} \in \tilde{\Delta}_e^{r_1} \\ 0 & \text{otherwise} \end{cases} .$$

Obviously, $\tilde{f}(i(k)) = f(k)$ for every $k \in \Delta_e^{r_1}$.

Now let h be a positive r_1-integrable harmonic function with a representing measure $\tilde{\mu}_h$ relative to r_1. Then $\tilde{\mu}_h = i_{r_1}(\mu_h)$ for some $\mu_h \in \mathcal{M}_+(\Delta_e^{r_1})$. Since f is assumed to be bounded on compact subsets of $\Delta_e^{r_1}$, the measure $f\mu_h$ is a positive regular Borel measure on $\Delta_e^{r_1}$. Consequently.

$$i_{r_1}(f\mu_h) = \tilde{f}\tilde{\mu}_h$$

is a positive regular Borel measure on $\tilde{\Delta}_e$ which is supported by $\tilde{\Delta}_e^{r_1} \subset \tilde{\Delta}_e^{r_2}$. For every $k \in \tilde{k} \in \tilde{\Delta}_e^{r_2}$,

$$\frac{k(x)}{\int k \, dr_1} = \frac{k(x)}{\int k \, dr_2} \; \frac{\int k \, dr_2}{\int k \, dr_1} ,$$

hence $\tilde{K}_x^{r_1} = \tilde{K}_x^{r_2} \cdot \tilde{f}$ for every $x \in X$. This implies

$$h(x) = \int \tilde{K}_x^{r_1} \, d\tilde{\mu}_h = \int \tilde{K}_x^{r_2} \, d(\tilde{f}\tilde{\mu}_h) \quad (x \in X),$$

but $\tilde{f}\tilde{\mu}_h$ is not necessarily a finite measure. According to theorem (2.11), $\tilde{f}\tilde{\mu}_h$ is the representing measure of h relative to r_2 iff

$\tilde{f}$ is $\tilde{\mu}_h$-integrable. By Fubini's theorem, the relation

$$\int \tilde{K}_x^{r_2}(\tilde{k})\,dr_2(x) = 1 \qquad (\tilde{k} \in \tilde{\Delta}_e^{r_2})$$

implies that the above statements are equivalent to the r_2-integrability of h.

<u>Special case</u>: If there exists a constant $\alpha > 0$ such that

$$(*) \qquad \frac{1}{\alpha}\cdot\int h\,dr_1 \leq \int h\,dr_2 \leq \alpha\int h\,dr_1$$

for every $h \in {}_+\mathcal{H}_X$, then the representing measures $\tilde{\mu}_h$ and $\tilde{\mu}'_h$ of a positive r_1-integrable harmonic function h relative to r_1, r_2 respectively satisfy

$$\tilde{\mu}_h = \tilde{f}\tilde{\mu}_h ,$$

where $\frac{1}{\alpha} \leq \tilde{f} \leq \alpha$ $\tilde{\mu}_h$-almost everywhere. Condition (*) is satisfied, if $(X,\mathcal{H})$ is a connected Brelot space and if r_1, r_2 are Dirac measures (cf. Harnack's inequality). ⌟

For the remainder of this section assume: <u>$(X,\mathcal{H})$ is a strong harmonic space</u>.

On ${}_+\mathcal{H}_X$, the topology of uniform convergence on compact subsets of X coincides with the 'T-topology' introduced by G.MOKOBODZKI (cf. M.SIEVEKING [21], p. 18, Eigenschaft 8). Hence we obtain from chapter I in [21] the following results:

<u>2.15:</u> For every $\tilde{k} \in \tilde{\Delta}_e$,

$$\underline{\tau_{\tilde{k}}} := \{E \subset X : R_k^{\complement E} \neq k \quad \text{for} \quad k \in i^{-1}(\tilde{k})\}$$

defines a filter on X. Being finer than the filter of the complements of relatively compact subsets of X, $\tau_{\tilde{k}}$ has empty adherence for every $\tilde{k} \in \tilde{\Delta}_e$. We call $\tau_{\tilde{k}}$ the <u>co-fine neighborhood filter of $\tilde{k}$</u>. We denote by

$$\underset{x \to \tilde{k}}{\text{f-lim}}\; u(x)$$

the limit of a function $u: X \longrightarrow \overline{\mathbb{R}}$ along the filter $\tau_{\tilde{k}}$.

With the aid of these filters, we consider a Dirichlet problem for the Martin boundary under the following assumption:

2.16 HYPOTHESIS: r is a normed reference measure and the constant function 1 is harmonic.

2.17: Denote by $\tilde{\mu}_1$ the representing measure of the harmonic function 1. $\tilde{\mu}_1$ is a probability measure which is supported by $\tilde{\Lambda}^r_e$. The same is true for every measure $\tilde{K}^r_x\tilde{\mu}_1$ $(x \in X)$.

2.18 For a numerical function $\tilde{f}$ on $\tilde{\Delta}_e$, define the set $\mathcal{O}_{\tilde{f}}$ of all upper functions of $\tilde{f}$ by

$$\mathcal{O}_{\tilde{f}} := \{v \in \mathcal{H}^*_X : v \text{ is bounded below and } \underset{x \to \tilde{k}}{\text{f-lim inf}}\ v(x) \geq \tilde{f}(\tilde{k})\ \tilde{\mu}_1\text{- a.e.}\}.$$

Every r-integrable upper function $v \in \mathcal{O}_{\tilde{f}}$ is superharmonic, since the support of r is contained in the absorbing set $\overline{\{x \in X : v(x) < \infty\}}$ (cf. (1.1) and [1], 1.4.2). Let for any $x \in X$

$$\overline{H}_{\tilde{f}}(x) := \inf \{v(x) : v \in \mathcal{O}_{\tilde{f}}\}$$

and

$$\underline{H}_{\tilde{f}}(x) := -\overline{H}_{-\tilde{f}}(x) .$$

The definition of upper functions implies immediately the equality $\overline{H}_{\tilde{f}} = \overline{H}_{\tilde{g}}$ for functions $\tilde{f}$, $\tilde{g}$ which agree $\tilde{\mu}_1$-almost everywhere on the Martin boundary $\tilde{\Delta}_e$.

2.19: We call a numerical function $\tilde{f}$ on $\tilde{\Delta}_e$ (r-)resolutive, iff

$$\inf \{\int v\, dr : v \in \mathcal{O}_{\tilde{f}}\} = -\inf \{\int w\, dr : w \in \mathcal{O}_{-\tilde{f}}\} < \infty .$$

For a resolutive function $\tilde{f}$, there exist r-integrable, hence superharmonic functions in $\mathcal{O}_{\tilde{f}}$ and $\mathcal{O}_{-\tilde{f}}$, whence $\overline{H}_{\tilde{f}}$ and $\underline{H}_{\tilde{f}}$ are harmonic and r-integrable; moreover, $\underline{H}_{\tilde{f}} \leq \overline{H}_{\tilde{f}}$ and $\int(\overline{H}_{\tilde{f}}-\underline{H}_{\tilde{f}})dr = 0$ imply $\overline{H}_{\tilde{f}} = \underline{H}_{\tilde{f}}$. If $\tilde{f}$ is resolutive, we call $H_{\tilde{f}} := \overline{H}_{\tilde{f}} = \underline{H}_{\tilde{f}}$ the solution of the Dirichlet problem for $\tilde{f}$ (relative to r).

The following propositions hold:

2.20 PROPOSITION: Every $\tilde{\mu}_1$-integrable function $\tilde{f}$ on $\tilde{\Delta}_e$ satisfies

$$\underset{x\to\tilde{k}}{\text{f-lim}} \int \tilde{f}\, \tilde{K}_x^r \, d\tilde{\mu}_1 = \tilde{f}(\tilde{k}) \quad \tilde{\mu}_1\text{- a.e. on } \tilde{\Delta}_e.$$

2.21 PROPOSITION: For every positive superharmonic function s on X there is a $\tilde{\mu}_1$-integrable function $\tilde{f}$ on $\tilde{\Delta}_e$ such that

$$\underset{x\to\tilde{k}}{\text{f-lim}}\; s(x) = \tilde{f}(\tilde{k}) \quad \tilde{\mu}_1\text{- a.e. on } \tilde{\Delta}_e \;;$$

moreover, if s is a potential, then $\tilde{f} = 0$.

2.22 PROPOSITION: For every positive bounded harmonic function h on X, we have $h = H_{\tilde{h}}$ where $\tilde{h}(\tilde{k}) = \underset{x\to\tilde{k}}{\text{f-lim}}\; h(x)$ $\tilde{\mu}_1$- a.e. on $\tilde{\Delta}_e$.

2.23 PROPOSITION: Every $\tilde{\mu}_1$-integrable function $\tilde{f}$ on $\tilde{\Delta}_e$ is resolutive and the solution is given by

$$H_{\tilde{f}}(x) = \int \tilde{K}_x^r \, \tilde{f} \, d\tilde{\mu}_1 \qquad (x \in X).$$

Conversely, every resolutive boundary function is $\tilde{\mu}_1$-integrable.

PROOFS: With the exception of the second part of (2.23), all these propositions (2.20) - (2.23) can be proved as in [21].

Let $\tilde{f}$ be a resolutive function on $\tilde{\Delta}_e$. Then any upper function of $\tilde{f}$ has a superharmonic r-integrable minorant u in $\mathcal{O}_{\tilde{f}}$. Especially u is bounded below, and according to (2.21) there exists a $\tilde{\mu}_1$-integrable function $\tilde{u}$ on $\tilde{\Delta}_e$ such that

$$\tilde{u} = \text{f-lim}\; u \quad \tilde{\mu}_1\text{- a.e. on } \tilde{\Delta}_e.$$

Obviously, u is an upper function of $\tilde{u}$, hence

$$\int^* \tilde{f}\, \tilde{K}_x^r \, d\tilde{\mu}_1 \leq \int \tilde{u}\, \tilde{K}_x^r \, d\tilde{\mu}_1 = H_{\tilde{u}}(x) \leq u(x) \qquad (x \in X).$$

Since this inequality holds for all such functions u we obtain

$$\int^* \tilde{f}\, \tilde{K}_x^r \, d\tilde{\mu}_1 \leq \overline{H}_{\tilde{f}}(x) \qquad (x \in X)$$

and likewise

$$\int_* \tilde{f}\, \tilde{K}^r_x \, d\tilde{\mu}_1 \geq \underline{H}_{\tilde{f}}(x) \qquad (x \in X) .$$

The assumed resolutivity of $\tilde{f}$ implies

$$H_{\tilde{f}}(x) = \int_* \tilde{f}\, \tilde{K}^r_x \, d\tilde{\mu}_1 = \int^* \tilde{f}\, \tilde{K}^r_x \, d\tilde{\mu}_1 \qquad (x \in X) .$$

By Fubini's theorem and by the r-integrability of the Dirichlet solution $H_{\tilde{f}}$ we obtain the $\tilde{\mu}_1$-integrability of $\tilde{f}$, since for every $\tilde{k} \in \tilde{\Delta}^r_e$

$$\int \tilde{K}^r_x(\tilde{k})\, dr(x) = 1. \qquad \_/$$

<u>2.24 REMARK</u>: Let $\tilde{f}$ be a $\tilde{\mu}_1$-integrable function on the Martin boundary. According to (2.23) and (2.11), the harmonic function $H_{\tilde{f}}$ has the representing measure $\tilde{f}\tilde{\mu}_1$(relative to r).

In the following sections we shall give a characterization of those harmonic functions whose representing measures are absolutely continuous with respect to $\tilde{\mu}_1$.

<u>2.25 EXAMPLE</u>: We apply the results of this section to the example (2.3) of the solutions of the heat equation in the upper half plane. The notations are those introduced in (2.3).

a) It is shown, that $\Delta^r_e = \{h_a : a \in \mathbb{R}\}$ is the set of all functions h which lie on an extreme ray of ${}_+\mathcal{H}_X$ such that $\int h\, dr = 1$. Moreover, every extreme ray of ${}_+\mathcal{H}_X$ intersects Δ^r_e. Hence the map

$$a \longrightarrow \{c \cdot h_a : c > 0\}$$

defines a bijection from $\mathbb{R}$ onto $\tilde{\Delta}_e$. For $\tilde{h}_a \in \tilde{\Delta}_e$, a base of neighborhoods is given by the sets

$$U_{K,\varepsilon} := \{\tilde{h}_b : b \in \mathbb{R}, \sup_{(x,t)\in K} |h_a(x,t) - c \cdot h_b(x,t)| \leq \varepsilon$$

$$\text{for some } c > 0\},$$

where $K \subset X$ is compact and $\varepsilon > 0$. By the definition of the h_a's, $\tilde{\Delta}_e$ and Δ^r_e are homeomorphic to $\mathbb{R}$, i.e. the Martin boundary of $(X,\mathcal{H})$ and the topological boundary of X in $\mathbb{R}^2$ coincide.

According to (2.10), the kernel which appears in the integral representation (2.11) is given by

$$\tilde{K}^r_{(x,t)}(a) = c_a t^{-\frac{1}{2}} \exp(-\frac{(x-a)^2}{4t}) \qquad (a \in \mathbb{R})$$

where

$$c_a = \begin{cases} \frac{a^2}{1-\exp(-a^2)} & \text{if } a \neq 0 \\ 1 & \text{if } a = 0 \end{cases} .$$

b) In (2.15), the filter of co-fine neighborhoods of a point a of the Martin boundary has been introduced by

$$\tau_a : = \{E \subset X : R_{h_a}^{\complement E} \neq h_a\} .$$

If $\mathfrak{U}(a,o)$ denotes the filter of neighborhoods of (a,o) in the Euclidean space $\mathbb{R}^2$, then $X \cap \mathfrak{U}(a,o)$ is coarser than τ_a : Indeed, for $U \in X \cap \mathfrak{U}(a,o)$, the function h_a is bounded on $X \setminus U$, hence $R_{h_a}^{\complement U}$ is bounded on X and consequently inequal to the unbounded function h_a, i.e. $U \in \tau_a$.

Moreover, $\underline{\tau_a \text{ is strictly finer than } X \cap \mathfrak{U}(a,o)}$.

Indeed, let

$$E_a : = \{(x,t) \in X : h_a(x,t) > \frac{1}{c_a}\} .$$

Then $h_a \leqq \frac{1}{c_a}$ on $X \setminus E_a$, hence $E_a \in \tau_a$. More concrete, we obtain

$$E_a = \{(x,t) \in X : (x-a)^2 < -2t \log t\},$$

whence $E_a \notin X \cap \mathfrak{U}(a,o)$, since the sequence of points

$$(x_n, t_n) : = (a + \sqrt{2 \log(n^{1/n})}, \frac{1}{n}) \in \mathbb{R} \times \mathbb{R}_+^* \qquad (n \in \mathbb{N})$$

converges to (a,o) in the Euclidean topology of $\mathbb{R}^2$ but $(x_n, t_n) \notin E_a$ for every $n \in \mathbb{N}$.

c) Now we compute the representing measure $\tilde{\mu}_1$ of the harmonic function 1 (relative to r). Let λ be the Lebesgue measure on $\mathbb{R}$. Then

$$\int \tilde{K}^r_{(x,t)}(a)\ \frac{1-\exp(-a^2)}{2\cdot\sqrt{\pi}\cdot a^2}\ d\lambda(a)$$

$$= \frac{1}{2\sqrt{\pi}} \int t^{-1/2} \exp(-\frac{(x-a)^2}{4t}) d\lambda(a) = 1$$

and by partial integration

$$\int \frac{1-\exp(-a^2)}{2\cdot\sqrt{\pi}\cdot a^2}\ d\lambda(a) = \frac{1}{\sqrt{\pi}} \int \exp(-a^2) d\lambda(a) = 1 .$$

According to theorem (2.11), the representing measure of 1 must be given by the probability measure

$$\tilde{\mu}_1 = \frac{1-\exp(-a^2)}{2\cdot\sqrt{\pi}\cdot a^2} \cdot \lambda \quad ,$$

i.e. $\tilde{\mu}_1$ has a strictly positive continuous density with respect to the Lebesgue measure λ.

The representing measure of a function lying on an extreme ray of ${}_+\mathcal{H}_X$ is a multiple of the Dirac measure of some point in $\mathbb{R}$, especially it is singular with respect to $\tilde{\mu}_1$.

3. Uniform integrability

In this chapter we generalize to the given set up the notion of uniform integrability, which has been introduced into potential theory by J.L.DOOB.

In [7], M.BRELOT developed a concept of uniform integrability for harmonic spaces which satisfy the axioms of H.BAUER with a weaker axiom of convergence (K_1 instead of K_D).

Our definition differs from the definition in [7] by replacing Dirac measures by one reference measure. In this way, we obtain generalizations of theorems of J.L.DOOB for harmonic spaces which satisfy the axioms of H.BAUER [1]. If we consider the special case of a connected Brelot space and if we choose a Dirac measure as a reference measure, these generalizations reduce exactly to known theorems. In general, these results contain some relations between uniform integrability of harmonic functions and their behavior near the Martin boundary. These relations will be deepened in the later chapters.

Let $(X_i, \Sigma_i, \mu_i)_{i\in I}$ be a family of measure spaces such that $\sup_i \mu(X_i) < \infty$. For every $i \in I$ let f_i be a Σ_i-measurable function on X_i. Then, $(f_i)_{i\in I}$ is called uniformly integrable relative to $(\mu_i)_{i\in I}$, iff $\lim_{a\to\infty} \int_{[|f_i|\geq a]} |f_i| d\mu_i = 0$ uniformly in $i \in I$. According to P.A.MEYER ([17], II §2), we have

<u>3.1 THEOREM</u>: The following statements are equivalent:

i) (f_i) is uniformly integrable with respect to (μ_i).

ii) $\sup_i \int |f_i| d\mu_i < \infty$ and for any family (ξ_i) of maps $\xi_i: \mathbb{R}_+\setminus\{0\} \longrightarrow \Sigma_i$ such that $\mu_i(\xi_i(a)) < a$ for every $a > 0$, we have

$$\lim_{a\to 0} (\sup_i \int_{\xi_i(a)} |f_i| d\mu_i) = 0 .$$

iii) There exists an increasing function Φ on $\mathbb{R}_+$ such that $\Phi(0) = 0$, $\lim_{t\to\infty} \frac{\Phi(t)}{t} = \infty$, and $\sup_i \int \Phi(|f_i|) d\mu_i < \infty$. Moreover, Φ can be chosen to be convex.

In the sequel, we always shall assume:

<u>3.2 HYPOTHESIS</u>: $(X,\mathcal{H})$ is a harmonic space, the constant function 1 is superharmonic, r is a normed reference measure, and $(U_i)_{i\in\mathbb{N}}$ is an exhaustion of X.

For $i \in \mathbb{N}$, we denote by μ_i the Radon measure on X defined by

$$\int f\, d\mu_i := \iint f(y) d\mu_x^{U_{i+1}}(y)\ 1_{\overline{U}_i}(x) dr(x) \quad (f \in \mathcal{K}(X)).$$

μ_i is supported by U_{i+1}^* and because of (3.2), we always have

$\mu_i(1) \leq 1$ $(i \in \mathbb{N})$.

<u>3.3 DEFINITION</u>: A Borel measurable function f on X is called <u>r-uniformly integrable</u> (in abbreviation: <u>r-u.i.</u>) iff the constant family (f) is uniformly integrable relative to the family $(\mu_i)_{i \in \mathbb{N}}$ of Radon measures on X.

From (3.1) we obtain immediately

<u>3.4 PROPOSITION</u>: A Borel measurable function f on X is r-uniformly integrable iff one of the following conditions is satisfied:

i) $\lim\limits_{a\to\infty} \int_{\overline{U}_i} \int_{[|f|\geq a]} |f(y)|\,d\mu_x^{U_{i+1}}(y)\,dr(x) = 0$ uniformly in $i \in \mathbb{N}$.

ii) $\sup\limits_i \int_{\overline{U}_i} \int |f|\,d\mu_x^{U_{i+1}}\,dr(x) < \infty$ and for any choice of Borel subsets $E_i(\varepsilon)$ of X such that $\int_{\overline{U}_i} \int_{E_i(\varepsilon)} d\mu_x^{U_{i+1}}\,dr(x) < \varepsilon$ for all $i \in \mathbb{N}$, we have

$$\lim_{\varepsilon\to 0} \sup_i \int_{\overline{U}_i} \int_{E_i} |f|\,d\mu_x^{U_{i+1}}\,dr(x) = 0.$$

iii) There is a real (convex) increasing function Φ on $\mathbb{R}_+$ such that $\Phi(0)=0$, $\lim\limits_{t\to\infty} \frac{\Phi(t)}{t} = \infty$, and $\sup\limits_i \int_{\overline{U}_i} \int \Phi(|f|)\,d\mu_x^{U_{i+1}}\,dr(x) < \infty$.

<u>REMARK</u>: If $|f|$ is subharmonic in X and if U is an open relatively compact subset of X, then the smallest harmonic majorant of $\mathrm{rest}_U |f|$ is given by

$$x \longrightarrow \int |f|\,d\mu_x^U \qquad (x \in U).$$

Hence, the r-uniform integrability for such a function f is independent of the special choice of the exhaustion (U_i).

3.5 LEMMA: Let u be a harmonic function such that

$$\sup_i \int_{\overline{U}_i} \int |u| d\mu_x^{U_{i+1}} dr(x) < \infty .$$

Then u is r-integrable. Moreover, the function

$$x \longrightarrow \lim_{i\to\infty} \int u^+ \, d\mu_x^{U_{i+1}} \qquad (x \in X)$$

is the smallest r-integrable positive harmonic majorant of u.

This holds especially if u is r-uniformly integrable.

PROOF: Applying lemma (1.5) to the subharmonic functions $u^+ = \sup(u,0)$ and $u^- = \sup(-u,0)$ we obtain the first part of the assertion. (3.4.ii) implies the second part. _/

3.6 THEOREM (DOOB): For any harmonic function u on X, the following conditions are equivalent:

i) u is r-uniformly integrable.

ii) For any $\varepsilon > 0$, there are r-integrable lower bounded superharmonic functions v_1, v_2 on X such that

$$-v_1 \leq u \leq v_2, \int (v_1 + v_2) \, dr \leq \varepsilon.$$

Moreover, in this case v_1 and v_2 can be chosen to be harmonic.

PROOF: a) First suppose, u is an r-u.i. function. Because of (3.2), for any real number $a > 0$, the function $u_a := \sup(u,-a)$ is lower bounded, subharmonic, and satisfies

$$-a \leq u_a \leq |u|,$$

hence u_a is r-u.i.. According to (3.4.ii) and (1.5), u_a has a smallest r-integrable harmonic majorant h_a; obviously

$$u \leq u_a \leq h_a.$$

Now for any $i \in \mathbb{N}$ and $x \in U_{i+1}$ the following estimates hold:

$$0 \leq \int u_a \, d\mu_x^{U_{i+1}} - u(x) = \int_{[u\leq -a]} (u_a - u) d\mu_x^{U_{i+1}}$$

$$\leq \int_{[|u|\geq a]} |u| d\mu_x^{U_{i+1}} ,$$

hence

$$0 \leq \int_{\overline{U}_i} \left(\int u_a \, d\mu_x^{U_{i+1}} - u(x)\right) dr(x)$$

$$\leq \int_{\overline{U}_i} \int_{[|u| \geq a]} |u| \, d\mu_x^{U_{i+1}} \, dr(x) \quad .$$

Let $\varepsilon > 0$. The assumed r-uniform integrability of u implies the existence of a real number $a > 0$ such that the last term is smaller than $\frac{\varepsilon}{2}$. For such an a, we obtain by (1.5)

$$0 \leq \int (h_a - u) \, dr$$

$$= \lim_i \int_{\overline{U}_i} \left(\int u_a \, d\mu_x^{U_{i+1}} - u(x)\right) dr(x) \leq \frac{\varepsilon}{2} \, .$$

Let $v_2 := h_a$. Then v_2 is a lower bounded harmonic function such that $u \leq v_2$ and $\int (v_2 - u) \, dr \leq \frac{\varepsilon}{2}$.

Applying the same arguments to $-u$, we obtain a lower bounded harmonic function v_1 such that $-u \leq v_1$ and $\int (v_1 - (-u)) \, dr \leq \frac{\varepsilon}{2}$. Then v_1 and v_2 do the job.

b) Let $\varepsilon > 0$ be arbitrary. Assume that v_1 and v_2 are superharmonic functions having the properties of ii). Then there is a real $K > 0$ such that $-v_1 \leq K$, $-v_2 \leq K$, hence

$$|u| \leq v_1 + v_2 + K \, .$$

For any $i \in \mathbb{N}$, $x \in U_{i+1}$, and $a \geqslant 0$, the positivity of $v_1 + v_2$ implies

$$(*) \qquad \int_{[|u| \geq a]} |u| \, d\mu_x^{U_{i+1}} \leq v_1(x) + v_2(x) + K \cdot \int_{[|u| \geq a]} d\mu_x^{U_{i+1}} \, ,$$

hence

$$\int_{\overline{U}_i} \int_{[|u| \geq a]} |u| \, d\mu_x^{U_{i+1}} \, dr(x) \leq \varepsilon + K =: K_o .$$

Especially, for $a = 0$ we obtain

$$\int_{\overline{U}_i} \int |u| \, d\mu_x^{U_{i+1}} \, dr(x) \leq K_o < \infty \qquad (i \in \mathbb{N}) \quad .$$

This yields for any $a > 0$

$$\int_{\overline{U}_i} \int_{[|u|\geq a]} a \, d\mu_x^{U_{i+1}} \, dr(x) \leq \int_{\overline{U}_i} \int_{[|u|\geq a]} |u| d\mu_x^{U_{i+1}} \, dr(x) \leq K_o$$

and consequently

$$\int_{\overline{U}_i} \int_{[|u|\geq a]} d\mu_x^{U_{i+1}} \, dr(x) \leq \frac{K_o}{a} \quad (i \in \mathbb{N}) .$$

If a is sufficiently large, the inequality (*) implies for any $i \in \mathbb{N}$

$$\int_{\overline{U}_i} \int_{[|u|\geq a]} |u| d\mu_x^{U_{i+1}} \, dr(x) \leq \varepsilon + \frac{K \cdot K_o}{a} \leq 2\varepsilon .$$

This is exactly the r-uniform integrability condition for u. _/

<u>3.7 COROLLARY</u>: A positive harmonic function u is r-uniformly integrable if and only if u is r-integrable and quasi-bounded, i.e. there exists an isotone sequence (u_n) of bounded positive harmonic functions increasing to u.

<u>PROOF</u>: a) Suppose that u is r-u.i.. For every $n \in \mathbb{N}$, the positive superharmonic function $\inf(u,n)$ has a greatest harmonic majorant u_n. Define $h: = \sup u_n$. Since $h \leq u$, u is harmonic and quasi-bounded. We show $h = u$. Let $n \in \mathbb{N}$ and choose for $\varepsilon = \frac{1}{n}$ and u functions $v_{1,n}$ and $v_{2,n}$ as in the preceding theorem. If $k = k(n)$ is sufficiently large we obtain

$$-v_{1,n} \leq u_k \leq h \leq u \leq v_{2,n}$$

and

$$0 \leq \int (u-h)dr \leq \int (v_{2,n} + v_{1,n})dr \leq \frac{1}{n} .$$

Since this inequality holds for every $n \in \mathbb{N}$ we obtain $u = h$.

b) The converse is an immediate consequence of (3.6) Indeed, put in (3.6) $v_2 = u$ and $v_1 = u_n$ for a suitable $n \in \mathbb{N}$ such that $\int (u-u_n)dr$ is sufficiently small. _/

<u>3.8 THEOREM (DOOB)</u>: Suppose $(X,\mathcal{H})$ is a strong harmonic space such that the constant function 1 is harmonic. Let

$\tilde{\mu}_1$ be the representing measure of 1 relative to r and let u be a positive harmonic function. Then the following statements are equivalent:

i) u is r-uniformly integrable;

ii) there exists an r-resolutive positive function $\tilde{f}$ on the Martin boundary such that u is the solution for the Dirichlet problem for $\tilde{f}$. Moreover, $\tilde{f}$ = f-lim u $\tilde{\mu}_1$-a.e. on $\tilde{\Delta}_e$.

PROOF: a) Assume u r-u.i.; according to (3.7), there is a sequence (u_n) of bounded positive harmonic functions increasing to u. According to (2.21), there exist positive $\tilde{\mu}_1$-integrable functions $\tilde{f}_n$, $\tilde{f}$ on $\tilde{\Delta}_e$ such that

$$\tilde{f}_n = \text{f-lim } \tilde{f}_n, \quad \tilde{f} = \text{f-lim } f \quad \tilde{\mu}_1\text{-a.e.} \quad (n \in \mathbb{N}).$$

We may assume the functions $\tilde{f}_n$ to be bounded and to increase to $\tilde{f}$. Then (2.22) and (2.23) imply

$$u_n(x) = H_{\tilde{f}_n}(x) = \int \tilde{K}_x^r \tilde{f}_n \, d\tilde{\mu}_1 \quad (x \in X, \quad n \in \mathbb{N}) ,$$

whence by Lebesgue's convergence theorem and (2.23)

$$u(x) = \int \tilde{K}_x^r \tilde{f} \, d\tilde{\mu}_1 = H_{\tilde{f}}(x) \quad (x \in X),$$

i.e. u is the solution of the boundary function $\tilde{f}$ relative to r.

b) Conversely, let $\tilde{f}$ be a resolutive function on $\tilde{\Delta}_e$. Then $u := H_{\tilde{f}}$ is the r-integrable solution for $\tilde{f}$. For $n \in \mathbb{N}$, define $\tilde{f}_n := \inf(\tilde{f},n)$. Then we obtain increasing sequences $(\tilde{f}_n)$ and $(H_{\tilde{f}_n})$ converging pointwise to $\tilde{f}$, $H_{\tilde{f}}$ respectively. Obviously, $H_{\tilde{f}_n}$ is bounded for every $n \in \mathbb{N}$, hence $u = \sup H_{\tilde{f}_n}$ is quasi-bounded and consequently r-u.i.. ⌟

Let g be an r-uniformly integrable function. Obviously, any Borel measurable function f such that $|f| \leq g$ is r-uniformly integrable, whence (3.8) yields immediately the following statement:

3.9 COROLLARY: Let $\tilde{g}$ be a positive resolutive function on the Martin boundary and let f be a harmonic function such that $0 \leq f \leq H_{\tilde{g}}$. Then $f = H_{\tilde{f}}$ where the $\tilde{\mu}_1$-integrable function $\tilde{f}$ is defined by

$$\tilde{f} = \text{f-lim } f \qquad \tilde{\mu}_1\text{-a.e. on } \tilde{\Delta}_e .$$

3.10 REMARK: Let h be a positive r-integrable harmonic function. If the representing measure $\tilde{\nu}$ of h is singular with respect to $\tilde{\mu}_1$, then

$$\text{f-lim } h = 0 \qquad \tilde{\mu}_1\text{-a.e. on } \tilde{\Delta}_e.$$

PROOF: According to (2.21), there exists a positive $\tilde{\mu}_1$-integrable function $\tilde{h}$ such that

$$\text{f-lim } h = \tilde{h} \qquad \tilde{\mu}_1\text{-a.e. on } \tilde{\Delta}_e.$$

Hence h is an upper function of $\tilde{h}$ (cf. 2.18), and we obtain $H_{\tilde{h}} \leq h$. Now let $\tilde{\nu}_1$ be the representing measure of $h - H_{\tilde{h}}$. Then

$$h = H_{\tilde{h}} + (h-H_{\tilde{h}})$$

implies

$$\tilde{\nu} = \tilde{h}\tilde{\mu}_1 + \tilde{\nu}_1$$

for the corresponding representing measures. Since $\tilde{\nu}$ is singular with respect to $\tilde{\mu}_1$, we obtain the desired result

$$\tilde{h} = 0 \qquad \tilde{\mu}_1\text{-a.e. on } \tilde{\Delta}_e.$$

4. $\mathcal{H}^\Phi$-spaces of harmonic functions

As in the preceding chapter we require the following conditions on the harmonic space $(X,\mathcal{H})$:

i) the constant function 1 is superharmonic,

ii) r is a normed reference measure,

iii) (U_i) is an exhaustion of X.

We denote by $_{\mathbb{C}}\mathcal{H}_X := \{u + iv : u,v \in \mathcal{H}_X\}$ the set of complex-valued harmonic functions and by $_{+*}\mathcal{H}_X$ the set of positive subharmonic functions on X. Following [16], we introduce some sets of such functions whose behavior near the Martin boundary will be studied in the next chapter.

4.1 DEFINITION: A strictly increasing convex function Φ on $\mathbb{R}_+$ such that $\Phi(0) = 0$, is called isovex. Moreover, if $\lim_{t\to\infty} \frac{\Phi(t)}{t} = \infty$, then Φ is called strongly isovex.
For any isovex function Φ we have $0 < \lim_{t\to\infty} \frac{\Phi(t)}{t} \leq \infty$, since $0 < t_o \leq t$ implies

$$\Phi(t_o) = \Phi\left(\frac{t_o}{t}\cdot t + \left(1 - \frac{t_o}{t}\right)\cdot 0\right) \leq \frac{t_o}{t}\,\Phi(t) ,$$

hence the map $t \longrightarrow \frac{\Phi(t)}{t}$ $(t \in \mathbb{R}_+^*)$ is isotone. Especially, $\lim_{t\to\infty} \Phi(t) = \infty$ for any isovex function Φ.

4.2 DEFINITIONS: Let Φ be a positive function on $\mathbb{R}_+$ and define $\underset{*}{\mathcal{H}}^{\Phi}(r)$ to be the set of all $s \in {}_{+}\underset{*}{\mathcal{H}}_X$ such that

i) $\Phi(s)$ is subharmonic,

ii) $\Phi(s)$ has an r-integrable harmonic majorant.

According to (1.7), for any $f \in {}_{\mathbb{C}}\mathcal{H}_X$ the function $|f|$ is subharmonic. We define

$$\mathcal{H}^{\Phi}(r) := \{f \in {}_{\mathbb{C}}\mathcal{H}_X : |f| \in \underset{*}{\mathcal{H}}^{\Phi}(r)\}\ ^{(*)} .$$

For $f \in \mathcal{H}^{\Phi} \cup \underset{*}{\mathcal{H}}^{\Phi}$, we denote by ${}_{\Phi}\{f\}$ the smallest harmonic majorant of $\Phi(|f|)$.

For any real $p > 0$, let $\mathcal{H}^p := \mathcal{H}^{\Phi_p}$ and $\underset{*}{\mathcal{H}}^p := \underset{*}{\mathcal{H}}^{\Phi_p}$ where Φ_p is defined by $\Phi_p(t) := t^p (t \geq 0)$.
Let $\mathcal{H}^{\infty}, \underset{*}{\mathcal{H}}^{\infty}$ be the set of bounded functions in ${}_{\mathbb{C}}\mathcal{H}_X$, ${}_{+}\underset{*}{\mathcal{H}}_X$ respectively.

4.3 REMARKS: 1) For $p = 1$ $(p > 1)$ the function

(*) Since we shall consider only the fixed reference measure r, there will be no confusion if we write $\mathcal{H}^{\Phi}$ or $\underset{*}{\mathcal{H}}^{\Phi}$ instead of $\mathcal{H}^{\Phi}(r)$ or $\underset{*}{\mathcal{H}}^{\Phi}(r)$.

$\Phi_p: t \longrightarrow t^p$ is isovex (strongly isovex).

2) If Φ is an isovex function and if $f \in {}_{\complement}\mathcal{H}_X \cup {}_{+*}\mathcal{H}_X$ then $\Phi(|f|)$ is subharmonic according to (1.6) and (1.7).

3) Suppose Φ is a positive function on $\mathbb{R}_+$ and let $f \in \mathcal{H}^\Phi \cup \underset{*}{\mathcal{H}}^\Phi$. (1.5) implies

$$\int_\Phi \{f\} dr = \sup_i \int_{\overline{U}_i} \int \Phi(|f|) d\mu_x^{U_{i+1}} dr(x) < \infty.$$

Immediately from (1.5) we receive the following useful characterization of functions in $\mathcal{H}^\Phi \cup \underset{*}{\mathcal{H}}^\Phi$:

<u>4.4 PROPOSITION</u>: Let Φ be a positive function on $\mathbb{R}_+$ and let $f \in {}_{\complement}\mathcal{H}_X (f \in {}_{+*}\mathcal{H}_X)$ such that $\Phi(|f|)$ is subharmonic. Then $f \in \mathcal{H}^\Phi (f \in \underset{*}{\mathcal{H}}^\Phi)$ if and only if there is a real $M \geq 0$ such that

$$\int_{\overline{U}_i} \int \Phi(|f|) d\mu_x^{U_{i+1}} dr(x) \leq M \quad \text{for all} \quad i \in \mathbb{N}.$$

<u>4.5 REMARK</u>: Assume Φ strongly isovex. (3.4.iii) implies that every function $f \in \mathcal{H}^\Phi \cup \underset{*}{\mathcal{H}}^\Phi$ is r-uniformly integrable.

The following properties are an easy consequence of the definitions and (4.4).

<u>4.6 COROLLARY</u>: Let Φ, Φ_1 be positive functions on $\mathbb{R}_+$. Then

i) $f \in \mathcal{H}^\Phi$, $\alpha, \beta \in \mathbb{R}_+$, $\Phi_0 := \alpha\Phi + \beta$, $\Phi_0(|f|) \in {}_{+*}\mathcal{H}_X \Rightarrow f \in \mathcal{H}^{\Phi_0}$.

ii) $f \in \mathcal{H}^\infty$, Φ locally bounded, $\Phi(|f|) \in {}_{+*}\mathcal{H}_X \Rightarrow f \in \mathcal{H}^\Phi$.

iii) $f \in \mathcal{H}^\Phi$, $0 \leq \Phi_1, \leq \Phi$, $\Phi_1(|f|) \in {}_{+*}\mathcal{H}_X \Rightarrow f \in \mathcal{H}^{\Phi_1}$.

iv) $f \in \mathcal{H}^\Phi$, $\liminf\limits_{t \to \infty} \frac{\Phi(t)}{t} > 0 \Rightarrow f \in \mathcal{H}^1$.

These statements remain valid if $\mathcal{H}^\Phi$ is replaced by $\underset{*}{\mathcal{H}}^\Phi$.

<u>4.7 COROLLARY</u>: For $1 \leq p \leq q \leq \infty$, we have $\mathcal{H}^\infty \subset \mathcal{H}^q \subset \mathcal{H}^p \subset \mathcal{H}^1$ and $\underset{*}{\mathcal{H}}^\infty \subset \underset{*}{\mathcal{H}}^q \subset \underset{*}{\mathcal{H}}^p \subset \underset{*}{\mathcal{H}}^1$.

4.8 REMARK: If Φ is an isovex, not strongly isovex function, then $0 < \lim\limits_{t\to\infty} \frac{\Phi(t)}{t}$, (4.3.2), and (4.6.iii) imply

$$\mathcal{H}^{\Phi} = \mathcal{H}^1, \quad \mathcal{H}_*^{\Phi} = \mathcal{H}_*^1 .$$

4.9 PROPOSITION: for any isovex function Φ, every real function in $\mathcal{H}^{\Phi}$ is the difference of positive functions in $\mathcal{H}^{\Phi}$.

PROOF: Let $u \in \mathcal{H}^{\Phi}$, hence $u \in \mathcal{H}^1$ (cf. (4.6.IV)). According to (4.4) and (3.5), the function u_1 defined by

$$u_1(x) := \lim_i \int u^+ \, d\mu_x^{U_{i+1}} \qquad (x \in X) ,$$

is the smallest r-integrable harmonic majorant of u^+. Since

$$u(x) = \int u^+ \, d\mu_x^{U_{i+1}} - \int u^- \, d\mu_x^{U_{i+1}} \qquad (i \in \mathbb{N}, \quad x \in U_{i+1}) ,$$

it suffices to prove $u_1 \in \mathcal{H}^{\Phi}$.

First of all, the convexity of Φ implies the subharmonicity of $\Phi(u_1)$. Now let $i \in \mathbb{N}$, $x \in U_{i+1}$ and put $c := \int d\mu_x^{U_{i+1}}$. Obviously, $0 \leq c \leq 1$.

If $c = 0$, we have

$$\Phi\left(\int u^+ \, d\mu_x^{U_{i+1}}\right) = \Phi(0) = 0 \leq {}_{\Phi}\{u\}(x) ,$$

if $c > 0$, Jensen's inequality applied to the probability measure $\nu := \frac{1}{c} \cdot \mu_x^{U_{i+1}}$ and the convex function Φ yields

$$\begin{aligned}
\Phi\left(\int u^+ \, d\mu_x^{U_{i+1}}\right) &= \Phi\left(c \cdot \int u^+ \, d\nu + (1-c) \cdot 0\right) \\
&\leq c \cdot \Phi\left(\int u^+ \, d\nu\right) \leq c \cdot \int \Phi(u^+) \, d\nu \\
&\leq \int \Phi(|u|) \, d\mu_x^{U_{i+1}} \leq {}_{\Phi}\{u\}(x) .
\end{aligned}$$

Hence by the continuity of Φ we obtain in both cases

$$\Phi(u_1(x)) = \lim_i \Phi\left(\int u \, d\mu_x^{U_{i+1}}\right) \leq {}_{\Phi}\{u\}(x) \qquad (x \in X)$$

whence $u_1 \in \mathcal{H}^{\Phi}$.

$\lrcorner$

4.10 REMARK: Suppose Φ is an isovex function. For any $f \in {}_{\Phi}\mathcal{H}_X \cup {}_{+}\mathcal{H}_{*X}$, $\Phi(|f|)$ is subharmonic according to (4.3.2), hence condition i) in the definition (4.2) is always satisfied. $\mathcal{H}^{\Phi}$ is a convex set but in general, not a linear space. A sufficient condition on Φ such that $\mathcal{H}^{\Phi}$ is a vector space is given by

$$\limsup_{t \to \infty} \frac{\Phi(2t)}{\Phi(t)} < \infty$$

(cf. (4.4)). Especially, the sets $\mathcal{H}^{p}$ are vector spaces for any $1 \leq p \leq \infty$.

4.11 DEFINITIONS: For the remainder of this section suppose that Φ is an isovex function. We denote

$$\underline{\mathcal{H}_1^{\Phi}} := \{f \in {}_{\mathbb{C}}\mathcal{H}_X : \alpha f \in \mathcal{H}^{\Phi} \text{ for every } \alpha > 0\},$$

$$\underline{\mathcal{H}_m^{\Phi}} := \{f \in {}_{\mathbb{C}}\mathcal{H}_X : \alpha f \in \mathcal{H}^{\Phi} \text{ for some } \alpha > 0\}.$$

Obviously, $\mathcal{H}_1^{\Phi}$ is the largest vector space contained in $\mathcal{H}^{\Phi}$ and $\mathcal{H}_m^{\Phi}$ is the smallest vector space containing $\mathcal{H}^{\Phi}$.

For a complex-valued or extended real-valued function f on X and for $i \in \mathbb{N}$ we introduce the notion

$$\underline{{}_i\|f\|_{\Phi}} := \inf \{\tfrac{1}{k} : 0 \leq k < \infty, \int_{\overline{U}_i}\int \Phi(k\cdot|f|)\,d\mu_x^{U_{i+1}}\,dr(x) \leq 1\}.$$

If $f \in {}_{\mathbb{C}}\mathcal{H}_X \cup {}_{+}\mathcal{H}^{*}_X$, then $\Phi(|f|)$ is subharmonic, hence $({}_i\|f\|_{\Phi})_{i \in \mathbb{N}}$ is an increasing sequence. We define

$$\underline{\|f\|_{\Phi}} := \sup_i {}_i\|f\|_{\Phi}.$$

4.12 REMARK: For a complex-valued or extended real-valued function f on X let

$$A_i := \{\tfrac{1}{k} : k \geq 0, \int_{\overline{U}_i}\int \Phi(k|f|)\,d\mu_x^{U_{i+1}}\,dr(x) \leq 1\} \quad (i \in \mathbb{N}).$$

The isotony of Φ and B.Levi's theorem of convergence imply

$$A_i = [{}_i\|f\|_{\Phi}, \infty] \setminus \{0\} \qquad (i \in \mathbb{N}),$$

hence for $f \in {}_{\mathbb{C}}\mathcal{H}_X \cup {}_{+}\mathcal{H}^{*}_X$,

$$\|f\|_\Phi = \inf \{\tfrac{1}{k}: 0 \leq k < \infty, \sup_i \int_{\overline{U}_i}\int \Phi(k|f|)\,d\mu_x^{U_{i+1}}dr(x) \leq 1\}.$$

4.13 LEMMA: A harmonic function f is in $\mathcal{H}_m^\Phi$ if and only if $\|f\|_\Phi$ is finite.

PROOF: a) Let $f \in \mathcal{H}_m^\Phi$, i.e. $\alpha f \in \mathcal{H}^\Phi$ for some $\alpha > 0$. According to (4.4) there is a real number M_1 such that

$$\int_{\overline{U}_i}\int \Phi(\alpha|f|)\,d\mu_x^{U_{i+1}}dr(x) \leq M_1 \quad \text{for every } i \in \mathbb{N}.$$

If $M_2 := \sup(1, M_1)$, then the convexity of Φ implies

$$\begin{aligned}
&\int_{\overline{U}_i}\int \Phi(\tfrac{\alpha}{M_2}|f|)\,d\mu_x^{U_{i+1}}dr(x) \\
&= \int_{\overline{U}_i}\int \Phi(\tfrac{\alpha}{M_2}|f| + (1 - \tfrac{1}{M_2})\cdot 0)\,d\mu_x^{U_{i+1}}dr(x) \\
&\leq \frac{1}{M_2}\int_{\overline{U}_i}\int \Phi(\alpha|f|)\,d\mu_x^{U_{i+1}}dr(x) \\
&\leq \frac{M_1}{M_2} \leq 1 \qquad\qquad (i \in \mathbb{N}),
\end{aligned}$$

i.e. $\|f\|_\Phi = \sup_i {}_i\|f\|_\Phi \leq \frac{M_2}{\alpha} < \infty$.

b) Conversely, according to (4.12), $\|f\|_\Phi < M < \infty$ implies

$$\int_{\overline{U}_i}\int \Phi(\tfrac{1}{M}|f|)\,d\mu_x^{U_{i+1}}dr(x) \leq 1 \quad \text{for every } i \in \mathbb{N}$$

whence $\frac{1}{M} f \in \mathcal{H}^\Phi$. ⌟

4.14 PROPOSITION: For any $i \in \mathbb{N}$, ${}_i\|\cdot\|_\Phi$ is a semi-norm and $\|\cdot\|_\Phi$ is a norm on $\mathcal{H}_m^\Phi$.

PROOF: a) First we show, that for any $i \in \mathbb{N}$ the map

$$f \longrightarrow {}_i\|f\|_\Phi$$

defines a semi-norm on $\mathcal{H}_m^\Phi$.

Let $f \in \mathcal{H}_m^\Phi$ and $\lambda \in \mathbb{C}$, $k, k' \in \mathbb{R}_+$ such that $k = k'\cdot|\lambda|$. Then the inequality

$$\int_{\overline{U}_i}\int \Phi(k|f|)\,d\mu_x^{U_{i+1}}dr(x) \leq 1$$

holds if and only if

$$\int_{\overline{U}_i}\int \Phi(k'\cdot|\lambda f|)\,d\mu_x^{U_{i+1}}\,dr(x) \leq 1$$

holds, i.e. $|\lambda|\,{}_i\|f\|_\Phi = {}_i\|\lambda f\|_\Phi$.

Now let $f, g \in \mathcal{H}_m^\Phi$ and let $c_1, c_2 \in \mathbb{R}$ such that

$${}_i\|f\|_\Phi < c_1,\ {}_i\|g\|_\Phi < c_2.$$

The isovexity of Φ and (4.12) yield

$$\int_{\overline{U}_i}\int \Phi\left(\frac{|f+g|}{c_1+c_2}\right)d\mu_x^{U_{i+1}}\,dr(x)$$

$$\leq \int_{\overline{U}_i}\int \left[\frac{c_1}{c_1+c_2}\cdot\Phi\left(\frac{|f|}{c_1}\right) + \frac{c_2}{c_1+c_2}\cdot\Phi\left(\frac{|g|}{c_2}\right)\right]d\mu_x^{U_{i+1}}\,dr(x) \leq 1 ,$$

i.e. ${}_i\|f+g\|_\Phi \leq c_1 + c_2$. Actually, this inequality implies the subadditivity of ${}_i\|\cdot\|_\Phi$.

b) Since ${}_i\|\cdot\|_\Phi$ is a semi-norm for every $i \in \mathbb{N}$, $\|\cdot\|_\Phi = \sup_i {}_i\|\cdot\|_\Phi$ is a semi-norm on $\mathcal{H}_m^\Phi$.

Now let $f \in \mathcal{H}_m^\Phi$ such that $\|f\|_\Phi = 0$. Then

$$\int_{\overline{U}_i}\int \Phi(n|f|)\,d\mu_x^{U_{i+1}}\,dr(x) \leq 1 \quad (i \in \mathbb{N})$$

and consequently $nf \in \mathcal{H}^\Phi$ for every $n \in \mathbb{N}$ (cf. (4.4)). Hence $({}_\Phi\{nf\})_{n\in\mathbb{N}}$ is an increasing sequence of harmonic functions whose r-integrals are bounded by 1 (cf.(4.3.3)). According to (1.4) $h:= \sup_\Phi\{nf\}$ is a harmonic and therefore a finite majorant of $\Phi(n|f|)$ $(n \in \mathbb{N})$. Now $\lim_{t\to\infty}\Phi(t) = \infty$ implies the desired result $f = 0$. ⌟

We need the following lemma to conclude the completeness of the normed space $(\mathcal{H}_m^\Phi, \|\cdot\|_\Phi)$.

<u>4.15 LEMMA</u>: For any sequence (f_n) in $\mathcal{H}_m^\Phi$, the following statements are equivalent:

i) $\lim_{n\to\infty} \|f_n\|_\Phi = 0$;

ii) $\lim_{n\to\infty} \left[\sup_i \int_{\overline{U}_i}\int \Phi(a|f_n|)\,d\mu_x^{U_{i+1}}\,dr(x)\right] = 0$

for any real number $a > 0$.

PROOF: a) Assume (ii). Then for any $a > 0$ there is an $n(a) \in \mathbb{N}$ such that for every $n \geq n(a)$

$$\sup_i \int_{\overline{U}_i} \int \Phi(a|f_n|) d\mu_x^{U_{i+1}} dr(x) \leq 1$$

and consequently $\|f_n\|_\Phi \leq \frac{1}{a}$. This implies $\lim_{n\to\infty} \|f_n\|_\Phi = 0$.

b) Conversely, if $(f_n) \subset \mathscr{H}_m^\Phi$ converges in Φ-norm to zero, then there is an $n(a) \in \mathbb{N}$ such that

$$\|f_n\|_\Phi \leq \frac{1}{a} \quad \text{for every} \quad n \geq n(a).$$

If $\|f_n\|_\Phi > 0$, we obtain by the convexity of Φ

$$\Phi(a|f_n|) = \Phi(a \cdot \|f_n\|_\Phi \cdot \frac{|f_n|}{\|f_n\|_\Phi})$$

$$\leq a \cdot \|f_n\|_\Phi \cdot \Phi(\frac{|f_n|}{\|f_n\|_\Phi}) \qquad (n \geq n(a)).$$

This implies

$$\int_{\overline{U}_i} \int \Phi(a|f_n|) d\mu_x^{U_{i+1}} dr(x) \leq a \, \|f_n\|_\Phi \qquad (i \in \mathbb{N})$$

for every $n \geq n(a)$. Consequently, $\lim_{n\to\infty} \|f_n\|_\Phi = 0$ yields ii). __/

4.16 THEOREM: $\mathscr{H}_m^\Phi$ is a Banach space in which $\mathscr{H}_1^\Phi$ is a closed subspace.

PROOF: Because of (4.14), we only have to prove the completeness of $\mathscr{H}_m^\Phi$ and $\mathscr{H}_1^\Phi$.

a) Let (f_n) be a Cauchy sequence in $\mathscr{H}_m^\Phi$. (4.15) yields

$$\lim_{n,m\to\infty} (\sup_i \int_{\overline{U}_i} \int \Phi(|f_n - f_m|) d\mu_x^{U_{i+1}} dr(x) = 0 ,$$

hence $f_n - f_m \in \mathscr{H}^\Phi$ for $n, m \in \mathbb{N}$ sufficiently large. Moreover, (4.3.3) implies

$$\lim_{n,m\to\infty} \int {}_\Phi\{f_n - f_m\} dr = 0.$$

Therefore ${}_\Phi\{f_n - f_m\}$ converges uniformly on compact subsets of X to zero, if n,m tend to infinity (cf. (1.2)). Since Φ is continuous at zero and since

$$\Phi(|f_n-f_m|) \leq {}_{\Phi}\{f_n-f_m\} \qquad (n,m \in \mathbb{N}) ,$$

the sequence (f_n) converges locally uniformly to some harmonic function $f \in {}_{\mathbb{C}}\mathscr{H}_X$.

It remains to show that $f \in \mathscr{H}_m^{\Phi}$ and that the sequence (f_n) converges in Φ-norm to f. Obviously,

$$\|f_n\|_{\Phi} \leq \|f_{n_o}\|_{\Phi} + \|f_n-f_{n_o}\|_{\Phi} \qquad (n,n_o \in \mathbb{N}),$$

thus there is some constant $k > 0$ such that $\|f_n\|_{\Phi} \leq k$ for every $n \in \mathbb{N}$. By Fatou's lemma and by the continuity of Φ,

$$\int_{\overline{U}_i}\int \Phi(\tfrac{1}{k}|f_n|)d\mu_x^{U_{i+1}}dr(x) \leq 1 \qquad (i,n \in \mathbb{N})$$

implies

$$\int_{\overline{U}_i}\int \Phi(\tfrac{1}{k}|f|)d\mu_x^{U_{i+1}}dr(x) \leq 1 \qquad (i \in \mathbb{N})$$

and consequently, $f \in \mathscr{H}_m^{\Phi}$ (cf. (4.13)).

To prove the norm-convergence of f_n to f, choose $a > 0$ arbitrarily. According to (4.15) there is an $n(a) \in \mathbb{N}$ such that

$$\int_{\overline{U}_i}\int \Phi(a|f_n-f_m|)d\mu_x^{U_{i+1}}dr(x) \leq 1 \qquad (i \in \mathbb{N})$$

for all $n,m \geq n(a)$. Fatou's lemma implies

$$\int_{\overline{U}_i}\int \Phi(a|f_n-f|)d\mu_x^{U_{i+1}}dr(x) \leq 1 \qquad (i \in \mathbb{N})$$

for all $n \geq n(a)$. Hence ${}_i\|f_n-f\|_{\Phi} \leq \frac{1}{a}$ for any $i \in \mathbb{N}$. Since this holds for any $a > 0$, we obtain $\lim_{n\to\infty} \|f_n-f\|_{\Phi} = 0$.

b) Finally we prove that $(f_n) \subset \mathscr{H}_1^{\Phi}$ implies $f \in \mathscr{H}_1^{\Phi}$. Let $a > 0$. Then $\lim_{n\to\infty} \|f_n-f\|_{\Phi} = 0$ yields the existence of a number $k = k(a) \in \mathbb{N}$ such that

$$\sup_i \int_{\overline{U}_i}\int \Phi(2a|f-f_k|d\mu_x^{U_{i+1}}dr(x) \leq 1.$$

Now $f_k \in \mathscr{H}_1^{\Phi}$ implies $2a\, f_k \in \mathscr{H}_1^{\Phi}$, hence (cf.(4.4))

$$\sup_i \int_{\overline{U}_i} \int \Phi(2a|f_k|)\, d\mu_x^{U_{i+1}}\, dr(x) < \infty.$$

Then the isovexity of Φ gives rise to

$$\sup_i \int_{\overline{U}_i} \int \Phi(a|f|)\, d\mu_x^{U_{i+1}}\, dr(x)$$

$$\leq \sup_i \{\tfrac{1}{2}\int_{\overline{U}_i} \int \Phi(2a|f-f_k|) + \Phi(2a|f_k|)\, d\mu_x^{U_{i+1}}\, dr(x)\} < \infty.$$

According to (4.4), $af \in \mathcal{H}^\Phi$ for any $a > 0$, i.e. $f \in \mathcal{H}^\Phi_1$. _/

4.17 REMARKS: a) $\mathcal{H}^\infty$ endowed with the supremum norm is a Banach space, too.

b) In the case $\Phi = \Phi_p : t \longrightarrow t^p$ $(1 \leq p < \infty)$, ${}_i\|\cdot\|_{\Phi_p}$ is given by

$${}_i\|f\|_{\Phi_p} = \left(\int_{\overline{U}_i} \int |f|^p d\mu_x^{U_{i+1}} dr(x)\right)^{\frac{1}{p}} \qquad (i \in \mathbb{N})$$

and we have

$$\|f\|_{\Phi_p} = \left(\int_{\Phi_p} \{f\} dr\right)^{\frac{1}{p}}.$$

In the special case $p = 2$, we obtain a Hilbert space $\mathcal{H}^2$, since the Φ_2-norm on $\mathcal{H}^2$ is induced by the inner product

$$(f,g) \longrightarrow \lim_i \int_{\overline{U}_i} \int f\bar{g}\, d\mu_x^{U_{i+1}} dr(x)$$

(the existence of these limits follows from (4.9) and (4.3.3)).

5. Boundary properties of $\mathcal{H}^\Phi$-functions

We require the following conditions on the strong harmonic space $(X,\mathcal{H})$:

i) The constant functions are harmonic;

ii) r is a normed reference measure relative to $(X,\mathcal{H})$;

iii) $\tilde{\mu}_1$ is the representing measure of the constant function 1 relative to r;

iV) (U_i) is an exhaustion of X.

We shall examine the behavior of harmonic functions near the Martin boundary. We shall see that the concept of r-uniform integrability is essential. As a central result we shall prove in (5.5) that for any strongly isovex function Φ on $\mathbb{R}_+$, $\mathcal{H}^\Phi_m$ is isomorphic to $L^\Phi_m(\tilde{\mu}_1)$, where the latter set is a Banach space of classes of functions on the Martin boundary.

5.1: As usual, let $\mathcal{L}^1(\tilde{\mu}_1)$ $(\mathcal{L}^\infty(\tilde{\mu}_1))$ be the set of $\tilde{\mu}_1$-integrable ($\tilde{\mu}_1$-essentially bounded) complex-valued functions on the Martin boundary.

For any isovex function Φ on $\mathbb{R}_+$, we define

$$\mathcal{L}^\Phi(\tilde{\mu}_1) := \{\tilde{f} : \tilde{f} \text{ is } \tilde{\mu}_1\text{-measurable such that } \Phi(|\tilde{f}|) \in \mathcal{L}^1(\tilde{\mu}_1)\},$$

$$\mathcal{L}^\Phi_m(\tilde{\mu}_1) := \{\tilde{f} : \alpha\tilde{f} \in \mathcal{L}^\Phi(\tilde{\mu}_1) \text{ for some } \alpha > 0\}.$$

Since $\lim_{t\to\infty} \frac{\Phi(t)}{t} > 0$, we have always $\mathcal{L}^\Phi(\tilde{\mu}_1) \subset \mathcal{L}^1(\tilde{\mu}_1)$.

By $L^1(\tilde{\mu}_1)$, $L^\infty(\tilde{\mu}_1)$, $L^\Phi(\tilde{\mu}_1)$, and $L^\Phi_m(\tilde{\mu}_1)$ we denote the corresponding quotient spaces associated with the usual equivalence relation $\tilde{f} \sim \tilde{g} :\Longleftrightarrow \int |\tilde{f}-\tilde{g}|\,d\tilde{\mu}_1 = 0$. The corresponding equivalence class of a function $\tilde{f}$ will be denoted by $[\tilde{f}]$.

5.2 DEFINITION: Let $\tilde{f}$ be a numerical function on $\tilde{\Delta}_e$. We define

$$\|\tilde{f}\|_\Phi := \inf\{\tfrac{1}{k} : k \geq 0, \int \Phi(k|\tilde{f}|)\,d\tilde{\mu}_1 \leq 1\}.$$

If $\tilde{f} = \tilde{g}$ $\tilde{\mu}_1$-a.e. on $\tilde{\Delta}_e$, we have obviously $\|\tilde{f}\|_\Phi = \|\tilde{g}\|_\Phi$, hence, if $[\tilde{f}] \in L^\Phi(\tilde{\mu}_1)$ then the expression

$$\|[\tilde{f}]\|_\Phi := \|\tilde{f}\|_\Phi$$

is well defined.

5.3 REMARK: If Φ is an isovex function on $\mathbb{R}_+$ then exactly one of the following two statements holds:

i) $\lim_{t\to\infty} \frac{\Phi(t)}{t} < \infty$; in this case $\mathcal{H}^\Phi = \mathcal{H}^1$ and $\mathcal{H}^\Phi_* = \mathcal{H}^1_*$ (cf. (4.8));

ii) Φ is strongly isovex; in this case every function in $\mathcal{H}^{\Phi} \cup \mathcal{H}_{*}^{\Phi}$ is r-uniformly integrable according to (4.5).

5.4 LEMMA: Suppose Φ is an isovex function on $\mathbb{R}_+$. For any $f \in \mathcal{H}_m^{\Phi}$ there is a $\tilde{\mu}_1$-integrable function $\tilde{f} \in \mathcal{L}_m^{\Phi}(\tilde{\mu}_1)$ such that $\tilde{f} = f - \lim f$ $\tilde{\mu}_1$-a.e. on $\tilde{\Delta}_e$.

Moreover, if Φ is strongly isovex then $f = H_{\tilde{f}}$.

PROOF: It suffices to consider $f \in \mathcal{H}^{\Phi}$. According to (4.9), every $f \in \mathcal{H}^{\Phi}$ is a complex linear combination of positive functions in $\mathcal{H}^{\Phi}$. Therefore we may assume that $f \geq 0$. (2.21) implies the existence of a $\tilde{\mu}_1$-integrable function $\tilde{f}$ on $\tilde{\Delta}_e$ such that $\tilde{f} = f\text{-}\lim f$ $\tilde{\mu}_1$-a.e. on $\tilde{\Delta}_e$. Application of (2.21) to the smallest harmonic majorant of $\Phi(|f|)$ implies

$$\tilde{h}(\tilde{k}) = \underset{x \to \tilde{k}}{f\text{-}\lim}\ {}_{\Phi}\{f\}(x) \geq \tilde{f}(\tilde{k}) \quad \tilde{\mu}_1\text{-a.e. on } \tilde{\Delta}_e,$$

where $\tilde{h}$ is a $\tilde{\mu}_1$-integrable function on $\tilde{\Delta}_e$. We conclude $\tilde{f} \in \mathcal{L}^{\Phi}(\tilde{\mu}_1)$.

Moreover, if Φ is strongly isovex then, according to (4.5), f is r-uniformly integrable. Hence Doob's theorem (3.8) implies $H_{\tilde{f}} = f$.

5.5 THEOREM: For any strongly isovex function Φ on $\mathbb{R}_+$, the map

$$\mathcal{J}_{\Phi} : [\tilde{f}] \longrightarrow H_{\tilde{f}}$$

defines a linear isometry between $L_m^{\Phi}(\tilde{\mu}_1)$ and $\mathcal{H}_m^{\Phi}$ (and between $L^{\infty}(\tilde{\mu}_1)$ and $\mathcal{H}^{\infty}$).

Moreover, if $\tilde{f} \in \mathcal{L}^{\Phi}(\tilde{\mu}_1)$ then

$${}_{\Phi}\{H_{\tilde{f}}\} = H_{\Phi(|\tilde{f}|)}$$

and

$${}_{1}\{H_{\tilde{f}}\} = H_{|\tilde{f}|} \ .$$

PROOF:(*) a) It is known that $L_m^{\Phi} \subset L^1$ and $\mathcal{H}_m^{\Phi} \subset \mathcal{H}^1$.

(*) If no confusion will arise we shall write L^{Φ} instead of $L^{\Phi}(\tilde{\mu}_1)$ etc.

Let $[\tilde{f}] \in L^1$. Since the solution $H_{\tilde{f}}$ is independent of the special choice of $\tilde{f} \in [\tilde{f}]$, $\mathcal{J}_\Phi$ is a well defined linear map from L_m^Φ into $\mathcal{H}^1$.

b) Next we show $\mathcal{J}_\Phi(L_m^\Phi) \subset \mathcal{H}_m^\Phi$.

Since $\mathcal{J}_\Phi$ is linear, it suffices to prove that $H_{\tilde{f}} \in \mathcal{H}^\Phi$ for any $\tilde{f} \in \mathcal{L}^\Phi$. For $x \in X$, the measure $\tilde{\nu}_x := \tilde{K}_x^r \tilde{\mu}_1$ is a probability measure on the Martin boundary. Hence, if $\Phi(|\tilde{f}|) \in \mathcal{L}^1$ and $f := H_{\tilde{f}}$ then Jensen's inequality implies

$$\Phi(|f(x)|) = \Phi(|H_{\tilde{f}}(x)|) \leq \Phi(\int |\tilde{f}|\tilde{K}_x^r \, d\tilde{\mu}_1)$$

$$\leq \int \Phi(|\tilde{f}|)d\tilde{\nu}_x = H_{\Phi(|\tilde{f}|)}(x) \qquad (x \in X),$$

whence $\Phi(|f|)$ possesses an r-integrable harmonic majorant, i.e. $f \in \mathcal{H}^\Phi$.

c) According to (5.4), $\mathcal{J}_\Phi$ is a bijection between L_m^Φ and $\mathcal{H}_m^\Phi$. Since the constant functions are harmonic, the preceding results also hold for L^∞ and $\mathcal{H}^\infty$ (note $L^\infty \subset L^2$ and $\mathcal{H}^\infty \subset \mathcal{H}^2$).

d) From b) we deduce for $\tilde{f} \in \mathcal{L}^\Phi$

$$\Phi(|H_{\tilde{f}}|) \leq {}_\Phi\{H_{\tilde{f}}\} \leq H_{\Phi(|\tilde{f}|)}.$$

Then the continuity of Φ implies according to (2.20)

$$\text{f-lim}\ {}_\Phi\{H_{\tilde{f}}\} = \Phi(|\tilde{f}|) \quad \tilde{\mu}_1\text{-a.e.},$$

hence (cf. 3.9)

$$H_{\Phi(|\tilde{f}|)} = {}_\Phi\{H_{\tilde{f}}\}.$$

Similarily,

$$|H_{\tilde{f}}| = |\int \tilde{f}\, \tilde{K}_\cdot^r \, d\tilde{\mu}_1| \leq \int |\tilde{f}|\tilde{K}_\cdot^r \, d\tilde{\mu}_1 = H_{|\tilde{f}|}$$

implies

$$H_{|\tilde{f}|} = {}_1\{H_{\tilde{f}}\}.$$

e) It remains to show that $\mathcal{J}_\Phi$ preserves the norm.

Firstly, let $\tilde{f} \in \mathcal{L}^\Phi$. Then Fubini's theorem and d) imply

$$\int_\Phi\{H_{\tilde f}\}dr = \int H_{\Phi(|\tilde f|)}dr = \int\int \Phi(|\tilde f|)\tilde K_x^r\, d\tilde\mu_1 dr(x)$$
$$= \int(\int \tilde K_x^r\, dr(x))\ \Phi\ (|\tilde f|)d\tilde\mu_1 = \int \Phi(|\tilde f|d\tilde\mu_1,$$

whence by (4.3.3) and (4.12) we obtain

$$\|H_{\tilde f}\|_\Phi = \inf\ \{\tfrac{1}{k} : k \geq 0,\quad \int_\Phi\{H_{k|\tilde f|}\}dr \leq 1\}$$
$$= \inf\ \{\tfrac{1}{k} : k \geq 0,\quad \int \Phi(k|\tilde f|)d\tilde\mu_1 \leq 1\} = \|\tilde f\|_{\tilde\Phi}.$$

The homogenity of the norm implies the above equations for any $\tilde f \in \mathcal{L}_m^\Phi$.

Finally, if $\tilde f \in \mathcal{L}^\infty$ then

$$\|\tilde f\|_{\tilde\infty} = \sup\ \{k : \tilde\mu_1([|\tilde f| > k] > 0\}.$$

Since f-lim $H_{\tilde f} = \tilde f$ $\tilde\mu_1$-a.e. on $\tilde\Delta_e$, we obtain

$$\|H_{\tilde f}\|_\infty = \sup_{x\in X} |H_{\tilde f}(x)| \geq \|\tilde f\|_{\tilde\infty},$$

whence the assertion follows from

$$\|H_{\tilde f}\|_\infty = \sup_{x\in X} |\int \tilde f\ \tilde K_x^r\ d\tilde\mu_1|$$
$$\leq \|\tilde f\|_{\tilde\infty} \cdot \sup_{x\in X} \int \tilde K_x^r\ d\tilde\mu_1 = \|\tilde f\|_{\tilde\infty}.$$

For an isovex function Φ, which is not strongly isovex, we have $\mathcal{H}^\Phi = \mathcal{H}^1$. In this case, theorem (5.5) does not hold. The reason is that a function in $\mathcal{H}^1$ is not necessarily r-uniformly integrable (cf. (3.8) and (3.10)). Only the following version of the integral representation theorem (2.11) can be obtained:

<u>5.6 THEOREM</u>: A function $f \in {}_{\mathbb{C}}\mathcal{H}_X$ belongs to $\mathcal{H}^1$ if and only if there exists a bounded complex regular Borel measure $\tilde\mu_f$ on the Martin boundary such that $\tilde\mu_f$ is supported by $\tilde\Delta_e^r$ and

$$f(x) = \int \tilde K_x^r\ d\tilde\mu_f \quad \text{for all} \quad x \in X.$$

In this case, $\tilde\mu_f$ is uniquely determined (it is the representing measure of f)

5.7 REMARK: For any function $f \in \mathscr{H}^1$, according to (5.4) there exists a $\tilde{\mu}_1$-integrable function $\tilde{f}$ such that $\tilde{f} = \text{f-lim } f$ $\tilde{\mu}_1$-a.e. on $\tilde{\Delta}_e$. If $\tilde{\mu}_f$ denotes the representing measure of f we obtain:

i) $\tilde{\mu}_f = \tilde{f}\tilde{\mu}_1 + \tilde{\nu}$ where $\tilde{\nu}$ is singular with respect to $\tilde{\mu}_1$.

ii) $\tilde{\nu} = 0$ if and only if f is r-uniformly integrable.

PROOF: Firstly, without loss of generality, we may assume that $f \geq 0$. If

$$\tilde{\mu}_f = \tilde{g}\tilde{\mu}_1 + \tilde{\nu}$$

denotes the Lebesgue decomposition of $\tilde{\mu}_f$ with respect to $\tilde{\mu}_1$ then we have obviously

$$f(x) = H_{\tilde{g}}(x) + \int \tilde{K}_x^r \, d\tilde{\nu} \quad (x \in X).$$

Since $\tilde{\nu}$ is singular with respect to $\tilde{\mu}_1$, we obtain from (3.10)

$$0 = \text{f-lim}_{x \to \tilde{k}} \int \tilde{K}_x^r \, d\tilde{\nu} \quad \tilde{\mu}_1\text{-a.e.},$$

hence $\tilde{f} = \text{f-lim } H_{\tilde{g}} = \tilde{g}$ $\tilde{\mu}_1$-a.e. on $\tilde{\Delta}_e$.

The last assertion follows from (3.8), (5.4), and the characterization of r-uniformly integrable functions by isovex functions. $\_\!/$

5.8 COROLLARY: Let Φ be an isovex function on $\mathbb{R}_+$ and let $s \in \mathscr{H}_*^{\Phi}$.

i) There is a function $\tilde{s} \in \mathscr{L}^{\Phi}$ such that

$$\tilde{s} = \text{f-lim } s \quad \tilde{\mu}_1\text{-a.e. on } \tilde{\Delta}_e.$$

ii) If Φ is strongly isovex then

$$_1\{s\} = H_{\tilde{s}}, \quad _\Phi\{s\} = H_{\Phi(\tilde{s})}.$$

PROOF: Since $\mathscr{H}_*^{\Phi} \subset \mathscr{H}_*^1$, (4.6.iv) yields

$$s + p_1 = {}_1\{s\}$$

and

$$\Phi(s) + p_\Phi = {}_\Phi\{s\}$$

for some potentials p_1 and p_Φ on X. According to (2.21), every potential has the co-fine limit zero $\tilde{\mu}_1$-almost everywhere on the Martin boundary, hence (5.4) and (5.5) imply the assertions.

6. Applications

We assume the same conditions on the strong harmonic space $(X,\mathcal{H})$ as in chapter 5.

In the last chapters we transfered the $\mathcal{H}^p$-theory for Brelot spaces to harmonic spaces satisfying the axioms of H.BAUER [1]. Following L.LUMER-NAIM [16], we shall now establish in our set up a general Phragmen-Lindelöf principle and an F. and M.Riesz theorem. In doing so, the concept of 'strongly subharmonic functions' of L.GÅRDING, L.HÖRMANDER [10] will be as useful as in [16].

6.1 DEFINITION: A positive subharmonic function s on X is called strongly subharmonic iff there exists a convex strictly increasing function χ on $\mathbb{R}$ such that

i) $\lim\limits_{t\to\infty} \frac{\chi(t)}{t} = \infty, \quad \lim\limits_{t\to-\infty} \chi(t) = 0,$

ii) $\chi^{-1}(s)$ is subharmonic on X.

χ is called a determining function of s.

6.2 THEOREM: Suppose s is a strongly subharmonic function with determining function χ. Let λ be a positive r-uniformly integrable function on X such that $s \leq \chi \circ \lambda$. Then

i) there is a real function $\tilde{s}$ on $\tilde{\Delta}_e$ such that

$$\chi^{-1}(\tilde{s}) \in \mathcal{L}^1(\tilde{\mu}_1),$$

$$\tilde{s} = \text{f-lim } s \quad \tilde{\mu}_1\text{-a.e.} \quad \text{on } \tilde{\Delta}_e,$$

and

$$s \leq \chi \cdot H_{\chi^{-1}(\tilde{s})} ;$$

ii) if $\tilde{s} \in \mathcal{L}^{\Phi}(\tilde{\mu}_1)$ for some isovex function Φ then $s \leq H_{\tilde{s}}$ and $s \in \mathcal{H}_*^{\Phi}$. Moreover, if $\tilde{s} \in \mathcal{L}^{\infty}(\tilde{\mu})$

then $s \leq H_{\tilde{s}}$ and s is bounded.

PROOF: a) Obviously,

$$0 \leq s_o := \sup(\chi^{-1}(s), 0) \leq \lambda,$$

i.e. $s_o \in \mathcal{H}_*^{\Phi'}$ for some strongly isovex function Φ' on $\mathbb{R}_+$ (cf.(3.4.iii) and (4.4)). According to (5.8), there exists a real function $\tilde{s}_o \in \mathcal{L}^1(\tilde{\mu}_1)$ such that

$$\tilde{s}_o = \text{f-lim } s_o = \text{f-lim } H_{\tilde{s}_o} \quad \tilde{\mu}_1\text{-a.e.},$$

$H_{\tilde{s}_o}$ is the smallest harmonic majorant of s_o.

$\chi^{-1}(s) \leq s_o$ implies: $H_{\tilde{s}_o} - \chi^{-1}(s)$ is a positive superharmonic function on X, which has a boundary limit $\tilde{f} \in \mathcal{L}^1(\tilde{\mu}_1)$:

$$\tilde{f} = \text{f-lim } (H_{\tilde{s}_o} - \chi^{-1}(s)) \quad \tilde{\mu}_1\text{-a.e.}.$$

Now, if $\tilde{s} := \chi(\tilde{s}_o - \tilde{f})$ then $\chi^{-1}(\tilde{s}) = \tilde{s}_o - \tilde{f} \in \mathcal{L}^1(\tilde{\mu}_1)$. By the continuity of χ, χ^{-1} and by the finiteness of the limits involved, we obtain $\tilde{\mu}_1$-a.e. on the Martin boundary

$$\tilde{s} = \chi(\tilde{s}_o - \tilde{f}) = \chi(\text{f-lim}[s_o - H_{\tilde{s}_o} + \chi^{-1}(s)])$$

$$= \chi(\text{f-lim } \chi^{-1}(s)) = \text{f-lim } s\ ,$$

hence the first two assertions of i) are proved.

b) $\chi^{-1}(s) \leq s_o \leq H_{\tilde{s}_o}$ implies that there exists a smallest r-integrable harmonic majorant h of $\chi^{-1}(s)$. Then $h = \chi^{-1}(s) + p$ for some potential p on X and according to (2.21),

$$\chi^{-1}(\tilde{s}) = \text{f-lim } \chi^{-1}(s) = \text{f-lim } h \quad \tilde{\mu}_1\text{-a.e.}.$$

$H_{\tilde{s}_o}$ is the smallest harmonic majorant of s_o, thus $\chi^{-1}(s) \leq s_o$ implies $h \leq H_{\tilde{s}_o}$. Then $H_{\tilde{s}_o} - h$ is an upper function of $\tilde{s}_o - \chi^{-1}(\tilde{s})$, especially

$$H_{(\tilde{s}_o - \chi^{-1}(\tilde{s}))} \leq H_{\tilde{s}_o} - h$$

and consequently

$$h \leq H_{\chi^{-1}(\tilde{s})}.$$

Since h is a harmonic majorant of $\chi^{-1}(s)$, the desired inequality

$$s \leq \chi \circ H_{\chi^{-1}(\tilde{s})}$$

follows from the isotony of χ.

c) To finish the proof of the theorem, let $\tilde{s} \in \mathcal{L}^{\Phi}$ for some isovex function Φ on $\mathbb{R}_+$, i.e.

$$\Phi(\tilde{s}),\ \tilde{s} \in \mathcal{L}^1(\tilde{\mu}_1).$$

By applying Jensen's inequality we deduce from b)

$$s \leq H_{\tilde{s}}$$

and

$$0 \leq \Phi(s) \leq (\Phi\circ\chi)(H_{\chi^{-1}(\tilde{s})}) \leq H_{\Phi(\tilde{s})} \in \mathcal{H}^1,$$

hence (cf. (4.2))

$$s \in \mathcal{H}_*^{\Phi}.$$

Let $\tilde{s} \in \mathcal{L}^{\infty}(\tilde{\mu}_1) \subset \mathcal{L}^2(\tilde{\mu}_1)$. Then (5.8) yields

$$0 \leq s \leq {}_1\{s\} = H_{\tilde{s}};$$

especially s is bounded since the constant functions are harmonic. $\lrcorner$

<u>6.3 THEOREM</u>: Any strongly subharmonic function $s \in \mathcal{H}_*^1$ satisfies the assumptions of (6.2). Moreover, we have $\tilde{s} \in \mathcal{L}^1(\tilde{\mu}_1)$, and $H_{\tilde{s}}$ is the smallest harmonic majorant of s.

<u>PROOF</u>: Let χ be a determining function of s. Then $\Phi: \mathbb{R}_+ \longrightarrow \mathbb{R}_+$ defined by $\Phi(t) := \chi(t) - \chi(0)$ is strongly isovex. Let $s_o := \sup(\chi^{-1}(s), 0)$. Since $s \in \mathcal{H}_*^1$ and $0 \leq \Phi(s_o) \leq s$, we have $s_o \in \mathcal{H}_*^{\Phi}$. Especially, s_o is r-uniformly integrable and $s \leq \chi\circ s_o$, i.e. the assumptions of (6.2) are satisfied.

According to (5.8), $s \in \mathcal{H}_*^1$ implies $\tilde{s} \in \mathcal{L}^1(\tilde{\mu}_1)$, hence by (6.2.ii) we obtain

$$s \leq h \leq H_{\tilde{s}},$$

where h is the smallest harmonic majorant of s. Now f-lim s = $\tilde{s}$ = f-lim $H_{\tilde{s}}$ $\tilde{\mu}_1$-a.e. and (3.9) give rise to $h = H_{\tilde{s}}$. $\lrcorner$

A harmonic function in $\mathcal{H}^1$ is not necessarily the solution of the Dirichlet problem for some boundary function (cf.(5.7)). But we can prove the following F. and M.Riesz theorem:

6.4 COROLLARY: Let $f \in \mathcal{H}^1$ such that $|f|$ is strongly subharmonic. Then $f = H_{\tilde{f}}$ for a $\tilde{\mu}_1$-integrable function $\tilde{f}$ on the Martin boundary, i.e. $\tilde{f}\tilde{\mu}_1$ is the representing measure of f. Moreover, we have $\tilde{f} = f\text{-}\lim f$ $\tilde{\mu}_1$-a.e. on $\tilde{\Delta}_e$.

PROOF: If $f \in \mathcal{H}^1$ then $|f| \in \mathcal{H}^1_*$, hence (6.3) implies $|f| \leq H_{|\tilde{f}|}$. Since $H_{|\tilde{f}|}$ is r-uniformly integrable, $|f|$ and consequently f are r-uniformly integrable, whence $f = H_{\tilde{f}}$ for the f-limit function $\tilde{f}$ (cf. (3.8), (3.4.ii), and (5.4)). $\_\!/$

6.5 EXAMPLES: 1) In the classical situation, (6.4) contains the F. and M.Riesz theorem:

Let $(X,\mathcal{H})$ be the Brelot space of the solutions of the Laplace equation in the unit disc X of the plane. Choose ε_0, the Dirac measure in zero, as a reference measure. The Martin boundary of this harmonic space is the unit sphere. The representing measure $\tilde{\mu}_1$ of the harmonic function 1 relative to ε_0 is given by the normed surface area on the unit sphere. Let f be a holomorphic function on X. Then f is harmonic on X, moreover, $|f|$ is strongly subharmonic with determining function $\chi := \exp$ (cf. W.RUDIN [20], theorem 17.3). According to (6.4), if $|f|$ has a harmonic majorant, i.e. if $f \in \mathcal{H}^1(\varepsilon_0)$, then f has a complex representing measure $\tilde{\mu}_f$ on the sphere such that $\tilde{\mu}_f$ is absolutely continuous with respect to $\tilde{\mu}_1$. This means

THEOREM (F. and M.RIESZ): Every holomorphic function in $\mathcal{H}^1(\varepsilon_0)$ has a representing measure which is absolutely continuous with respect to the surface area.

2) Now we return to the situation of a general strong harmonic space $(X,\mathcal{H})$. The theorems (6.2) and (6.3) may be seen as

versions of the Phragmen-Lindelöf principle:

Let h be any positive harmonic function on X. Then there exists a normed reference measure r on X such that h is r-integrable. We denote by $\tilde{\mu}_1$ the representing measure of the harmonic function 1 relative to this measure r.
If $p > 1$ then the map $\Phi_p: t \longrightarrow t^p$ is a strongly isovex function on $\mathbb{R}_+$. Hence the harmonicity of h implies that $h^{1/p}$ is r-uniformly integrable (cf. (3.4)). Thus we obtain by (6.2):

<u>THEOREM (PHRAGMEN-LINDELÖF principle)</u>: Let s be a positive subharmonic function such that log s is subharmonic. Further let h be a positive harmonic function such that s satisfies the growth condition

$$\text{(GC)} \qquad s \leq \exp(h^{1/p}) \quad \text{for some} \quad p > 1.$$

Then f-lim inf $s \leq 1$ $\tilde{\mu}_1$-a.e. on $\tilde{\Delta}_e$ implies $s \leq 1$.

Obviously, the quality of this theorem depends on the success whether or not it is possible to find a rapidly increasing harmonic function h.

In the special case when s is the modulus of a holomorphic function in a domain of the plane, we can find such theorems for example in W.RUDIN's book (cf. [20], (12.7) and (12.9)).

3) M.H.PROTTER and H.F.WEINBERGER proved in [18] (II §9 and III §6) theorems of Phragmen-Lindelöf type for functions which are subharmonic relative to some elliptic or parabolic differential equation. In the case of the heat equation, we shall now develop a concrete version of (6.5.2) which will be compared with a result of [18].

Suppose that $(X,\mathcal{H})$ is the strong harmonic space of the solutions of the heat equation in the strip $R^n \times]0,t_0[$ $(n \in \mathbb{N}, 0 < t_0 \leq \infty)$. According to M.SIEVEKING ([21], p. 59), the Martin boundary of $(X,\mathcal{H})$ can be identified with the set $\mathbb{R}^n \times \{0\}$, since every element of an extreme ray of ${}_+\mathcal{H}_X$ is a multiple of the unbounded function

$$h_a : (x,t) \longrightarrow t^{-n/2} \exp(-\frac{(x-a)^2}{4t}),\ (x,t) \in X,\ a \in \mathbb{R}^n.$$

Furthermore, exactly as in the case $n = 1$ we can show that the filter of co-fine neighborhoods of a Martin boundary point $(a,0) \in \mathbb{R}^n \times \{0\}$ is finer (even strictly finer) than the intersection of the Euclidean neighborhood filter of $(a,0)$ in $\mathbb{R}^{n+1}$ with X (cf. example (2.25.b)). Now we can state the following theorem:

THEOREM: Let $c, c_1, \ldots, c_n \in \mathbb{R}$ and let s be a subharmonic function on X such that

$$\text{(GC)'} \quad s(x,t) \leq \exp(\sum_{i=1}^{n} (c_i x_i + 2c_i^2 t) + c) \quad \text{on } X.$$

Let N be an at most denumberable subset of $\mathbb{R}^n$. If $\limsup_{(x,t)\to(z,o)} s(x,t) \leq 0$ for every $z \in \mathbb{R}^n \setminus N$ then $s(x,t) \leq 0$ for every $(x,t) \in \mathbb{R}^n \times]0,t_o[$.

PROOF: The function $h: X \longrightarrow \mathbb{R}$ defined by

$$h(x,t) := \exp(2 \sum_{i=1}^{n} (c_i x_i + 2c_i^2 t) + 2c)$$

is harmonic on X. Especially, h is r-integrable for a suitable normed reference measure r. If $\tilde{\mu}_1$ denotes the representing measure of the harmonic function 1 then the finite subsets of the Martin boundary have $\tilde{\mu}_1$-measure zero. Otherwise $\varepsilon_{\tilde{k}} \ll \tilde{\mu}_1$ for some $\tilde{k} \in \tilde{\Delta}_e$ and this would imply that the extremal function h_k is bounded. Consequently $N \times \{0\}$ has $\tilde{\mu}_1$-measure zero.

$h^{1/2}$ is r-uniformly integrable. The growth condition (GC)' implies $\exp(s) \leq \exp h^{1/2}$. By (1.6), exp s is a positive subharmonic function with subharmonic logarithm. Obviously, $s \leq 0$ iff $\exp s \leq 1$, hence the assertion of the theorem follows by the theorem in (6.5.2) if we recall that the co-fine neighborhood filters are finer than the Euclidean ones. ⌟

The growth condition (GC)' in the last theorem is more restrictive than

$$(GC)'' \quad \liminf_{\rho \to \infty} (\exp(-c\rho^2) \sup_{\substack{x^2=\rho^2 \\ o<t<t_o}} s(x,t)) \leq 0,$$

which is used in [18] (§6 theorem 10).

But if we replace (GC)' by (GC)", our theorem does not remain valid unless $N = \emptyset$.

This is easily seen by the following counter example:
The harmonic function $s: X \longrightarrow \mathbb{R}$ defined by

$$s(x,t): = t^{-n/2} \exp(- \frac{x^2}{4t})$$

satisfies

$$\liminf_{\rho \to \infty} (\exp(-\rho^2) \sup_{o<t<t_o} t^{-n/2} \exp(- \frac{\rho^2}{4t})) = 0$$

as well as

$$\lim_{(x,t)\to(z,o)} s(x,t) = 0 \quad \text{for every} \quad z \in \mathbb{R}^n \setminus \{0\};$$

but nevertheless $s(x,t) > 0$ for all $(x,t) \in \mathbb{R}^n \times]0,t_o[$.

Bibliography

H.BAUER [1] : Harmonische Räume und ihre Potentialtheorie. Lecture Notes in Mathematics 22. Springer-Verlag Berlin-Heidelberg-New York (1966).

[2] : Konvexität in topologischen Vektorräumen. Lecture notes. Hamburg (1964).

N.BOURBAKI :
[3] Topologie générale I,II, 4e édition.

[4] : Topologie générale IX, 2e édition.

[5] : Intégration I-IV, 2e édition.

[6] : Espaces vectoriels topologiques I,II, 2e édition. Hermann, Paris

M.BRELOT [7] : Intégrabilité uniforme. Séminaire Théorie du potentiel, Paris (1961/62).

G.CHOQUET [8] : Axiomatique des mesures maximales. C.R.Acad.Sc. de Paris (1962) p.37-39.

J.L.DOOB [9] : Probability methods applied to the first boundary value problem. Third Berkeley Symp. on Math.Statistics and Prob. 2 (1954/55) p.49-80.

L.GÅRDING, L.HÖRMANDER [10] : Strongly subharmonic functions. Math.Scand. 15 (1964) p.93-96.

K.GOWRISANKARAN [11] : Extreme harmonic functions and boundary value problems. Ann.Inst.Fourier 13/2 (1963) p.307-356.

[12] : Fatou-Naïm-Doob limit theorems in the axiomatic system of Brelot. Ann.Inst.Fourier 16/2 (1966) p.455-467.

K.HOFMANN [13] : Banach spaces of analytic functions. Prentice Hall (1962).

J.LEMBCKE [14] : Konservative Abbildungen und Fortsetzung regulärer Maße. Thesis, University of Erlangen-Nürnberg (1969).

L.LUMER-NAIM [15] : Fonctions sous-analytique, classes H^{Φ}, principe de Phragmen-Lindelöf. C.R.Acad.Sc. de Paris 262 (1962) p.1167.

[16] : H^p spaces of harmonic functions. Ann.Inst.Fourier 17/2 (1967) p.425-469.

P.A.MEYER [17] : Probabilités et potentiel. Hermann, Paris (1966).

M.H.PROTTER, H.F.WEINBERGER [18] : Maximum principles in differential equations. Prentice Hall (1967).

R.R.PHELPS [19] : Lectures on Choquet's theorem. Van Nostrand (1966).

W.RUDIN [20] : Real and complex analysis. Mc Graw Hill (1966).

M.SIEVEKING [21] : Integraldarstellung superharmonischer Funktionen mit Anwendung auf parabolische Differentialgleichungen. Lecture Notes in Mathematics 69 (1968). Springer-Verlag Berlin-Heidelberg-New York.

Symbols

Contents

APPROXIMATION OF CAPACITIES BY MEASURES

by

Bernd Anger

0. Introduction

In this paper we want to generalize results of V.STRASSEN and C.DELLACHERIE on the approximation of capacities by measures.

Strassen ([5], 4.3) proved that any capacity F which is monotone of order ∞ on a compact metric space is the pointwise minimum of all positive measures which are minorized by F and have the same total mass as F.

Dellacherie observed that any alternating capacity F of order 2 on a compact metric space is the pointwise supremum of all measures majorized by F.

Here we generalize these results on strongly superadditive (resp. strongly subadditive) capacities (i.e. monotone or alternating capacities of order 2, respectively) to l o c a l l y c o m p a c t Hausdorff spaces.

I want to thank Professor Maurice Sion for some useful discussions on this subject. The idea of using weak convergence in proposition (3.1) is due to him.

1. Preliminaries

Let E be a locally compact Hausdorff space and let $\mathfrak{K}(E)$ denote the set of all compact subsets of E.

1.1 DEFINITION: A capacity on E is a mapping

$$F : \mathfrak{K}(E) \longrightarrow \mathbb{R}_+$$

such that

(1) $F(\emptyset) = 0$,

(2) F is monotone increasing, i.e.

if $K_1, K_2 \in \mathfrak{K}(E)$ and $K_1 \subset K_2$ then $F(K_1) \leq F(K_2)$,

(3) F is continuous on the right, i.e.

for $K \in \mathfrak{K}(E)$ and $\varepsilon > 0$ there is an open set $U \supset K$ such that if $K' \in \mathfrak{K}(E)$ and $K \subset K' \subset U$ then $F(K') - F(K) < \varepsilon$.

A capacity is called strongly subadditive (resp. strongly super-additive) if

(4) for $K_1, K_2 \in \mathfrak{K}(E)$ we have

$F(K_1 \cup K_2) + F(K_1 \cap K_2) \leq F(K_1) + F(K_2)$ (resp.

$F(K_1 \cup K_2) + F(K_1 \cap K_2) \geq F(K_1) + F(K_2)$).

For a capacity F the inner capacity F_* and the outer capacity F^* are defined on the set $\mathfrak{P}(E)$ of all subsets of E by

(5) $F_*(X) = \sup \{F(K) : K \in \mathfrak{K}(E), K \subset X\}$ $(X \subset E)$

and by

(6) $F^*(X) = \inf \{F_*(U) : X \subset U \subset E, \ U \text{ open}\}$ $(X \subset E)$.

A subset X of E is called F-capacitable, if

(7) $F_*(X) = F^*(X)$.

1.2 REMARKS: (1) (1.1.1) is not an essential restriction:

If $F' : \mathfrak{K}(E) \longrightarrow \mathbb{R}_+$ satisfies (1.1.2) and (1.1.3), then the mapping

$$F := F' - F'(\emptyset)$$

is a capacity.

(2) For a capacity F the inner capacity and the outer capacity coincide with F on $\mathfrak{K}(E)$. Every $\mathfrak{K}$-analytic set in E,

which is contained in a countable union of compact subsets of E, is F-capacitable. In particular, every $\mathfrak{K}$-Borel set is F-capacitable (cf. [3], 9.3).

(3) It is well known (cf. [3], 9.7) that for a strongly subadditive capacity F, the outer capacity F* is a capacity of Choquet, i.e.

$$F^* : \mathcal{P}(E) \longrightarrow \mathbb{R}_+ \cup \{\infty\} =: \bar{\mathbb{R}}_+$$

is a monotone increasing mapping such that

(i) if $(X_n)_{n\in\mathbb{N}}$ is an increasing sequence of subsets of E, then

$$F^*(\bigcup_{n\in\mathbb{N}} X_n) = \sup_{n\in\mathbb{N}} F^*(X_n) ,$$

(ii) if $(X_n)_{n\in\mathbb{N}}$ is a decreasing sequence of compact subsets of E, then

$$F^*(\bigcap_{n\in\mathbb{N}} K_n) = \inf_{n\in\mathbb{N}} F^*(K_n) .$$

On the space $\mathbb{R}^E$ of all real-valued functions on E let '$\leq$' be the (partial) order defined by

$$f \leq g \quad \text{iff} \quad f(x) \leq g(x) \quad \text{for all} \quad x \in E.$$

For a subset $\mathcal{G}$ of $\mathbb{R}^E$ we define

$$\mathcal{G}_+ := \{g \in \mathcal{G} : g \geq 0\}$$

to be the set of positive elements of $\mathcal{G}$.

If f is any bounded function on E, set

$$\|f\| := \sup \{|f(x)| : x \in E\} .$$

Let $\mathcal{C}(E)$ denote the space of all real-valued continuous functions on E and let $\mathcal{K}(E)$ be the subspace of all functions $f \in \mathcal{C}(E)$ having compact support Supp f.

<u>1.3 DEFINITION:</u> For a capacity F define $\hat{F} : \mathcal{K}_+(E) \longrightarrow \mathbb{R}_+$ by

$$\hat{F}(f) := \int_{]0,\|f\|]} F(\{f \geq \alpha\}) \, \lambda(d\alpha),$$

where λ denotes the Lebesgue measure on $\mathbb{R}$. (Notice that the sets

$$\{f \geq \alpha\} : = \{x \in E : f(x) \geq \alpha\} \qquad (\alpha > 0)$$

are compact.)

1.4 PROPOSITION: For any capacity F, the functional $\hat{F}$ has the following properties:

(1) $\hat{F}$ is positively homogeneous, i.e.
if $f \in \mathcal{K}_+(E)$ and $\alpha \in \mathbb{R}_+$ then $\hat{F}(\alpha f) = \alpha \hat{F}(f)$.

(2) $\hat{F}$ is monotone increasing, i.e.
if $f_1, f_2 \in \mathcal{K}_+(E)$ and $f_1 \leq f_2$ then $\hat{F}(f_1) \leq \hat{F}(f_2)$.

(3) F is strongly subadditive (resp. strongly superadditive) iff $\hat{F}$ is subadditive (resp. superadditive), i.e.

if $f_1, f_2 \in \mathcal{K}_+(E)$ then $\hat{F}(f_1+f_2) \leq \hat{F}(f_1) + \hat{F}(f_2)$

(resp. if $f_1, f_2 \in \mathcal{K}_+(E)$ then $\hat{F}(f_1+f_2) \geq \hat{F}(f_1) + \hat{F}(f_2)$).

(4) For $X \subset E$, the indicator function of X is defined on E by

$$1_X(x): = \begin{cases} 1, & x \in X \\ 0, & x \notin X \end{cases} .$$

If $K \subset E$ is compact and if $f \in \mathcal{K}_+(E)$ such that

$$1_K \leq f \leq 1_X \quad ,$$

then

$$F(K) \leq \hat{F}(f) \leq F_*(X) \quad .$$

(5) If $K \subset E$ is compact and if $U \subset E$ is open, then

$$F(K) = \inf \{\hat{F}(f) : 1_K \leq f \in \mathcal{K}_+(E)\} \text{ and}$$
$$F_*(U) = \sup \{\hat{F}(f) : 1_U \geq f \in \mathcal{K}_+(E)\}.$$

(6) If F_1 and F_2 are capacities on E, then

$$F_1(K) \leq F_2(K) \quad \text{for all } K \in \mathfrak{K}(E) \qquad \text{iff}$$
$$\hat{F}_1(f) \leq \hat{F}_2(f) \quad \text{for all } f \in \mathcal{K}_+(E) \ .$$

We denote the (partial) order defined by (6) on the set of all capacities on E by '$\leq$'.

(3) was proved by Choquet ([2], 54.2), the other pro-

perties follow easily from the definition.

(2) shows that we can extend $\hat{F}$ to $\mathcal{C}_+(E)$ by the

1.5 DEFINITION: For a capacity F define

$\tilde{F} : \mathcal{C}_+(E) \longrightarrow \bar{\mathbb{R}}_+$ by

$$\tilde{F}(f) = \sup \{\hat{F}(g) : g \in \mathcal{K}_+(E),\ g \leq f\} .$$

1.6 PROPOSITION: For any capacity F, the functional $\tilde{F}$ has the following properties:

(1) $\tilde{F}$ is positively homogeneous.

(2) $\tilde{F}$ is monotone increasing.

(3) $\tilde{F}$ is subadditive (resp. superadditive) iff F is strongly subadditive (resp. strongly superadditive).

(4) If $f \in \mathcal{K}_+(E)$ and if $g \in \mathcal{C}_+(E)$ is bounded such that the restriction $g \mid \operatorname{Supp} f = \|g\|$, then

$$\tilde{F}(f + g) = \hat{F}(f) + \tilde{F}(g) .$$

(5) $\|F\| := \tilde{F}(1) = F_*(E)$.

PROOF: (1) and (2) obviously follow from (1.4.1,2).

To prove (3), by (1.4.3) it is sufficient to show: If $\hat{F}$ is subadditive (resp. superadditive), then $\tilde{F}$ is subadditive (resp. superadditive).

Let $\hat{F}$ be subadditive and let $f_1, f_2 \in \mathcal{C}_+(E)$, $g \in \mathcal{K}_+(E)$ such that $g \leq f_1 + f_2$. Choose $h \in \mathcal{K}_+(E)$ such that the restriction $h \mid \operatorname{Supp} g = \|h\| = 1$. Then for $i = 1,2$ the functions

$$g_i := h\, f_i \leq f_i$$

belong to $\mathcal{K}_+(E)$ and

$$g \leq g_1 + g_2 \leq f_1 + f_2 .$$

Therefore, we have

$$\hat{F}(g) \leq \hat{F}(g_1 + g_2) \leq \hat{F}(g_1) + \hat{F}(g_2) \leq \tilde{F}(f_1) + \tilde{F}(f_2),$$

hence

$$\tilde{F}(f_1 + f_2) \leq \tilde{F}(f_1) + \tilde{F}(f_2) .$$

If $\hat{F}$ is superadditive then obviously $\tilde{F}$ is superadditive.

To prove (4), by (1) it is sufficient to show that

$$\tilde{F}(f + g) = \hat{F}(f) + \tilde{F}(g)$$

for all $0 \neq f \in \mathcal{K}_+(E)$ and all $g \in \mathcal{C}_+(E)$ such that $g \mid \mathrm{Supp}\, f = \|g\| = 1$. Choose $h \in \mathcal{K}_+(E)$ such that $h \leq g$ and $h \mid \mathrm{Supp}\, f = 1$. Let us show that

$$\hat{F}(f + h) = \hat{F}(f) + \hat{F}(h) .$$

For $0 < \alpha \leq 1 = \|h\|$ we have

$$\{f + h \geq \alpha\} = \{h \geq \alpha\} .$$

For $1 < \alpha \leq \|f + h\| = \|f\| + 1$ we have

$$\{f + h \geq \alpha\} = \{f \geq \alpha - 1\} ,$$

therefore,

$$\hat{F}(f+h) = \int_{]0,\|f+h\|]} F(\{f+h \geq \alpha\})\ \lambda(d\alpha)$$

$$= \int_{]0,\|h\|]} F(\{h \geq \alpha\})\ \lambda(d\alpha) + \int_{]1,\|f\|+1]} F(\{f \geq \alpha - 1\})\ \lambda(d\alpha) = \hat{F}(h) + \hat{F}(f)$$

and hence

$$\tilde{F}(f+g) = \sup\ \{\hat{F}(k) : f+g \geq k \in \mathcal{K}_+(E)\}$$
$$= \sup\ \{\hat{F}(f+h) : g \geq h \in \mathcal{K}_+(E)\}$$
$$= \sup\ \{\hat{F}(f) + \hat{F}(h) : g \geq h \in \mathcal{K}_+(E),\ h \mid \mathrm{Supp}\, f = 1\}$$
$$= \hat{F}(f) + \tilde{F}(g).$$

(5) is an immediate consequence of (1.4.4).

2. Approximation of capacity functionals

We denote by $\mathcal{M}_+(E)$ the set of all positive Radon measures on E.

Suppose that $\mathcal{G}$ is a linear subspace of $\mathcal{C}(E)$. A numerical functional

$$p : \mathcal{G} \longrightarrow]-\infty, \infty]$$

will be called sublinear, if p is positively homogeneous and subadditive (we define $\alpha + \infty := \infty$ for $\alpha \in]-\infty, \infty]$, $\alpha\, \infty := \infty$ for $\alpha > 0$ and $0\, \infty := 0$). p is called a semi-norm, if p is real-valued, sublinear and symmetric (i.e. $p(g) = p(-g)$ for all $g \in \mathcal{G}$). A semi-norm is always positive.

2.1 PROPOSITION: Let F be a strongly subadditive capacity on E and let $K \subset U \subset E$ such that U is open and $K, \overline{U}$ are compact. There exist two sequences $(u_n)_{n\in\mathbb{N}}$ and $(e_n)_{n\in\mathbb{N}}$ in $\mathcal{K}_+(E)$ having the properties:

(1) $u_n \leq 1_U$ for all $n \in \mathbb{N}$ and $F_*(U) = \sup_{n\in\mathbb{N}} \hat{F}(u_n)$,

(2) $e_n \leq 1$ for all $n \in \mathbb{N}$ and $\|F\| = \sup_{n\in\mathbb{N}} \hat{F}(e_n)$,

(3) for $f \in \mathcal{K}_+(E)$ such that $\mathrm{Supp}\, f \subset K$ there exists a measure $\mu \in \mathcal{M}_+(E)$ such that $\mu \leq F$ and $\mu(h) = \hat{F}(h)$ for $h \in \{f, u_n, e_n : n \in \mathbb{N}\}$.

PROOF: By (1.4), the functional p defined on $\mathcal{K}(E)$ by

$$p(g) = \hat{F}(|g|)$$

is a semi-norm on $\mathcal{K}(E)$.

Let $I \subset \mathcal{K}_+(E)$ such that

(4) I is linearly independent and

(5) $p(\sum_{j\in J} \alpha_j\, j) = \sum_{j\in J} \alpha_j\, p(j)$ for all finite $J \subset I$ and for all $\alpha_j \geq 0$ $(j \in J)$.

Let $\mathcal{J}$ be the subspace of $\mathcal{K}(E)$ generated by I and let L_I be the linear form on $\mathcal{J}$ defined by

$$L_I(\sum_{j\in J} \alpha_j\, j) = \sum_{j\in J} \alpha_j\, p(j) \quad (\alpha_j \in \mathbb{R},\ j \in J \subset I,\ J \text{ finite}).$$

For $j \in I$ we have

$$L_I(j) = p(j) = \hat{F}(j).$$

If $\mathcal{V}$ denotes the set

$$\{v \in \mathcal{K}(E) : p(v) < 1\} ,$$

a theorem of H.Bauer and I.Namioka ([1], §1) states that L_I can be extended to a measure $\mu_I \in \mathcal{M}_+(E)$ having the property

$$|\mu_I(h)| \leq p(h) \quad \text{for all} \quad h \in \mathcal{K}(E)$$

$$\text{iff} \quad L(g) > -1 \quad \text{for all} \quad g \in \mathfrak{I} \cap (\mathcal{V} + \mathcal{K}_+(E)).$$

$\mathfrak{I} \cap (\mathcal{V} + \mathcal{K}_+(E))$ is the set of all elements of $\mathfrak{I}$ which are minorized by an element of $\mathcal{V}$. If $g \in \mathfrak{I} \cap (\mathcal{V} + \mathcal{K}_+(E))$, there exist an element $v \in \mathcal{V}$, a finite set $J \subset I$ and real numbers α_j, such that

$$v \leq g = \sum_{j \in J} \alpha_j \, j \; .$$

Put $J_+ := \{j \in J : \alpha_j \geq 0\}$, $J_- := J \setminus J_+$ and define $\sum_{j \in \emptyset} \alpha_j \, j := 0$. Then

$$0 \leq - \sum_{j \in J_-} \alpha_j \, j \leq \sum_{j \in J_+} \alpha_j \, j - v, \quad \text{hence by (1.4.2)}$$

$$p(- \sum_{j \in J_-} \alpha_j \, j) = \hat{F}(- \sum_{j \in J_-} \alpha_j \, j) \leq \hat{F}(\sum_{j \in J_+} \alpha_j j - v) = p(\sum_{j \in J_+} \alpha_j j - v).$$

It follows from (5) that

$$L_I(g) = \sum_{j \in J} \alpha_j \, p(j) = p(\sum_{j \in J_+} \alpha_j \, j) - p(- \sum_{j \in J_-} \alpha_j \, j)$$

$$\geq p(\sum_{j \in J_+} \alpha_j \, j) - p(\sum_{j \in J_+} \alpha_j j - v) \geq - p(-v) = -p(v) > -1.$$

So we proved that L_I can be extended to a measure $\mu_I \in \mathcal{M}_+(E)$ such that

$$|\mu_I(h)| \leq p(h) \qquad \text{for all} \quad h \in \mathcal{K}(E) \, ,$$

especially

$$\mu_I(h) \leq \hat{F}(h) \qquad \text{for all} \quad h \in \mathcal{K}_+(E) \quad \text{and}$$

$$\mu_I(j) = \hat{F}(j) \qquad \text{for all} \quad j \in I \, .$$

Let us now prove (2.1) by defining the set I. If U is compact, put $u_n = 1_U$; if E is compact, define $e_n = 1$ for all $n \in \mathbb{N}$.

If U is not compact, put $u_o := 0$, and define $e_o := 0$, if E is not compact.

For $n \in \mathbb{N}$ let

(1) $x_n \in U \setminus (K \cup \operatorname{Supp} u_{n-1})$ and by (1.4.5) $u_n \in \mathcal{K}_+(E)$ such that
$u_n \leq 1_U$, $u_n \mid K \cup \operatorname{Supp} u_{n-1} \cup \{x_n\} = 1$ and
$F_*(U) - \hat{F}(u_n) \leq \frac{1}{n}$ if $F_*(U) < \infty$,
$\hat{F}(u_n) \geq n$ if $F_*(U) = \infty$;

(2) $y_n \in \complement(\overline{U} \cup \operatorname{Supp} e_{n-1})$ and $e_n \in \mathcal{K}_+(E)$ such that
$e_n \leq 1$, $e_n \mid \overline{U} \cup \operatorname{Supp} e_{n-1} \cup \{y_n\} = 1$ and
$\|F\| - \hat{F}(e_n) \leq \frac{1}{n}$ if $\|F\| < \infty$,
$\hat{F}(e_n) \geq n$ if $\|F\| = \infty$.

Define

$$I := \{f, u_n, e_n : n \in \mathbb{N}\} \setminus \{0\}.$$

Then I is linearly independent. For finite $J \subset I$, $J \neq \{f\}$, there is $i \in J$ such that

$$i \leq 1 \quad \text{and} \quad i \mid \bigcup_{\substack{j \in J \\ j \neq i}} \operatorname{Supp} j = 1.$$

By (1.6.1,4) condition (5) is satisfied. The measure $\mu := \mu_I$ defined above has the desired property (3).

<u>2.2 COROLLARY</u>: Let F be a strongly subadditive capacity on E and let $f \in \mathcal{K}_+(E)$. Then

$$\hat{F}(f) = \max\{\mu(f) : \mu \in \mathcal{M}_+(E),\ \mu \leq F,\ \|\mu\| = \|F\|\}.$$

<u>PROOF</u>: Let $K := \operatorname{Supp} f$ and U be an open, relatively compact set, containing K. By (2.1), there is a sequence $(e_n)_{n \in \mathbb{N}}$ in $\mathcal{K}_+(E)$ and a measure $\mu \in \mathcal{M}_+(E)$ such that

$$\mu \leq F,\quad \mu(f) = \hat{F}(f) \text{ and } \mu(e_n) = \hat{F}(e_n) \text{ for all } n \in \mathbb{N}.$$

Therefore,

$$\begin{aligned} \|\mu\| &= \sup\{\mu(h) : h \in \mathcal{K}_+(E),\ h \leq 1\} \\ &\leq \sup\{\hat{F}(h) : h \in \mathcal{K}_+(E),\ h \leq 1\} = \check{F}(1) = \|F\| \\ &= \sup_{n \in \mathbb{N}} \hat{F}(e_n) = \sup_{n \in \mathbb{N}} \mu(e_n) \leq \|\mu\|, \text{ hence} \end{aligned}$$

$$\|\mu\| = \|F\|.$$

<u>2.3 PROPOSITION</u>: Let F be a strongly superadditive capacity on E and let $K \subset E$ be compact. There is a sequence $(k_n)_{n \in \mathbb{N}}$ in $\mathcal{K}_+(E)$ having the properties:

(1) $1_K \leq k_n$ for all $n \in \mathbb{N}$ and $F(K) = \inf_{n \in \mathbb{N}} \hat{F}(k_n)$,

(2) for $f \in \mathcal{K}_+(E)$ such that $\operatorname{Supp} f \subset K$ there exists a measure $\mu \in \mathcal{M}_+(E)$ such that

$\mu \geq F$, $\|\mu\| = \|F\|$ and $\mu(h) = \hat{F}(h)$ for $h \in \{f, k_n : n \in \mathbb{N}\}$.

<u>PROOF</u>: If $\|F\|$ is finite, let

$$\mathcal{G} := \{f + \alpha : f \in \mathcal{K}(E), \ \alpha \in \mathbb{R}\} ,$$

otherwise define $\mathcal{G} := \mathcal{K}(E)$.

By (1.6), the numerical functional

$$p : \mathcal{G} \longrightarrow]-\infty, \infty]$$

defined by

$$p(g) := \begin{cases} -\tilde{F}(g) , & g \in \mathcal{G}_+ \\ +\infty , & g \notin \mathcal{G}_+ \end{cases}$$

is sublinear.

Let I be a subset of $\mathcal{G}_+$ such that

(3) I is linearly independent and

(4) $p(\sum_{j \in J} \alpha_j j) = \sum_{j \in J} \alpha_j p(j)$ for all finite $J \subset I$ and for all $\alpha_j \geq 0$ $(j \in J)$.

Let $\mathcal{J}$ be the subspace of $\mathcal{G}$ generated by I, and let L_I be the linear form on $\mathcal{J}$ defined by

$$L_I(\sum_{j \in J} \alpha_j j) = \sum_{j \in J} \alpha_j p(j) \quad (\alpha_j \in \mathbb{R}, \ j \in J \subset I, \ J \text{ finite}).$$

For $j \in I$ we have

$$L_I(j) = p(j) = -\tilde{F}(j) \in \mathbb{R} .$$

Let us prove that L_I is majorized by p on $\mathcal{J}$:

For $g \in \mathcal{J}$ there are a finite set $J \subset I$ and real numbers α_j $(j \in J)$ such that

$$g = \sum_{j \in J} \alpha_j j .$$

Put $J_+ := \{j \in J : \alpha_j \geq 0\}$, $J_- := J \setminus J_+$. Then by (4)

$$L_I(\sum_{j \in J} \alpha_j j) = \sum_{j \in J} \alpha_j p(j) = p(\sum_{j \in J_+} \alpha_j j) - p(-\sum_{j \in J_-} \alpha_j j)$$

$$\leq p(\sum_{j \in J} \alpha_j j) .$$

By the theorem of Hahn-Banach, which is valid also for sublinear functionals with values in $]-\infty, \infty]$, there is an extension L of L_I to a linear form on $\mathcal{G}$ such that

$$L(g) \leq p(g) \qquad \text{for all} \quad g \in \mathcal{G} .$$

L is monotone decreasing since $g \in \mathcal{G}_+$ implies

$$L(g) \leq p(g) = -\tilde{F}(g) \leq 0 .$$

Hence, the restriction

$$\mu_I := -L \mid \mathcal{K}(E)$$

is a positive Radon measure on E.

We have

$$\mu_I(g) = -L(g) \geq -p(g) \qquad \text{for all} \quad g \in \mathcal{G},$$

especially

$$\mu_I(h) \geq -p(h) = \tilde{F}(h) = \hat{F}(h) \qquad \text{for all} \quad h \in \mathcal{K}_+(E),$$

hence $\mu_I \geq F$, and

$$\mu_I(i) = \tilde{F}(i) \qquad \text{for all} \quad i \in I.$$

Let us now prove (2.3) by defining the set I.

If K is open, set $k_n = 1_K$ for all $n \in \mathbb{N}$.

If K is not open, there is a sequence $(K_n)_{n \in \mathbb{N}}$ of compact sets in E such that for $n \in \mathbb{N}$

$$K \subset K_{n+1} \subsetneqq \mathring{K}_n \subset K_1 \neq E \quad \text{and} \quad F(K_n) - F(K) \leq \frac{1}{n} .$$

Let $x_n \in \mathring{K}_n \setminus K_{n+1}$ and choose $k_n \in \mathcal{K}_+(E)$ such that

$$k_n \leq 1, \; k_n \mid K_{n+1} \cup \{x_n\} = 1, \quad k_n \mid \complement \mathring{K}_n = 0.$$

We obtain

(1) $1_K \leq k_n \leq 1$ for all $n \in \mathbb{N}$ and by (1.4.4) $F(K) = \inf_{n \in \mathbb{N}} \hat{F}(k_n)$.

Define

$$I: = \{f, 1, k_n : n \in \mathbb{N}\} \setminus \{0\} \quad \text{if} \quad \|F\| < \infty ,$$

$$I: = \{ f, k_n : n \in \mathbb{N}\} \setminus \{0\} \quad \text{if} \quad \|F\| = \infty .$$

Then I is linearly independent. For finite $J \subset I$ there is $i \in I$ such that

$$i \leq 1 \quad \text{and} \quad i \mid \bigcup_{\substack{j \in J \\ j \neq i}} \operatorname{Supp} j = 1 .$$

By (1.6.1,4) condition (4) is satisfied.

The measure $\mu: = \mu_I$ defined above has the desired property (2), if we can show that $\|\mu\| = \|F\|$. This is clear if $\|F\| = \infty$, since $\|\mu\| \geq \|F\|$. Otherwise, we have $1 \in I$, hence $L_I(1) = - \tilde{F}(1)$ and therefore by (1.6.5),

$$\begin{aligned} \|\mu\| &= \sup \{\mu_I(h) : 0 \leq h \leq 1, \ h \in \mathcal{K}(E)\} \\ &= \sup \{-L(h) : 0 \leq h \leq 1, \ h \in \mathcal{K}(E)\} \leq -L(1) = -L_I(1) \\ &= \tilde{F}(1) = \|F\| \leq \|\mu\|, \ \text{hence} \end{aligned}$$

$\|F\| = \|\mu\|$.

<u>2.4 COROLLARY</u>: Let F be a strongly superadditive capacity on E and let $f \in \mathcal{K}_+(E)$. Then

$$\hat{F}(f) = \min \{\mu(f) : \mu \in \mathcal{M}_+(E), \ \mu \geq F, \ \|\mu\| = \|F\|\}.$$

3. Approximation of capacities

In the following, regular Borel measures on the σ-ring of all σ-bounded Borel sets and positive Radon measures are identified. For $\mu \in \mathcal{M}_+(E)$, the inner measure μ_* and the outer measure μ^* are defined on $\mathfrak{P}(E)$ as in (1.1). Observe that the restrictions of μ_* and μ^* on the Borel sets are measures extending μ, which coincide on open sets and on sets of finite outer measure.

We put

$$\mu(X) := \mu_*(X) = \mu^*(X)$$

for all μ-capacitable Borel sets X.

3.1 PROPOSITION: Let F be a strongly subadditive capacity on E. For any compact set $K \subset E$, we obtain

$$F(K) = \max \{\mu(K) : \mu \in \mathcal{M}_+(E),\ \mu \leq F,\ \|\mu\| = \|F\|\}.$$

PROOF: Let K be a compact subset of E. By (1.4.5) there is a decreasing sequence $(k_n)_{n\in\mathbb{N}}$ in $\mathcal{K}_+(E)$ such that

$$1_K \leq k_n \quad \text{and} \quad \hat{F}(k_n) - F(K) \leq \frac{1}{n} \quad \text{for all} \quad n \in \mathbb{N}.$$

Apply (2.1) to the compact set Supp k_1 and to an arbitrary open set U, such that $K \subset U$ and $\overline{U}$ is compact. There is a sequence $(e_n)_{n\in\mathbb{N}}$ in $\mathcal{K}_+(E)$ having the properties:

(1) $e_n \leq 1$ for all $n \in \mathbb{N}$,

(2) $\|F\| = \sup_{n\in\mathbb{N}} \hat{F}(e_n)$,

(3) for $f \in \mathcal{K}_+(E)$, $f \leq k_1$, there is a measure

$\mu \in M := \{\mu \in \mathcal{M}_+(E) : \mu \leq F,\ \mu(e_n) = \hat{F}(e_n)$ for all $n \in \mathbb{N}\}$

such that $\mu(f) = \hat{F}(f)$.

The set M is compact in the vague topology, since it is a closed subset of

$$\prod_{f\in\mathcal{K}(E)} [-F(|f|)\ ,\ F(|f|)].$$

For $n \in \mathbb{N}$ and $f \in \mathcal{K}_+(E)$ such that $1_K \leq f \leq k_n$ put

$$\tau_{n,f} = \{\mu \in M : \hat{F}(f) - \mu(f) \leq \tfrac{3}{n}\}.$$

Then $\tau_{n,f} \neq \emptyset$ since $f \leq k_1$ and hence by what was shown before there is a measure $\mu \in M$ such that $\mu(f) = \hat{F}(f)$.

For $i = 1,2$ let $n_i \in \mathbb{N}$ and $f_i \in \mathcal{K}_+(E)$ such that $1_K \leq f_i \leq k_{n_i}$. Define $m := 3 \sup(n_1, n_2)$ and $h := \inf(f_1, f_2, k_m)$. Then $h \in \mathcal{K}_+(E)$ and $1_K \leq h \leq k_m$. We have

$$\tau_{m,h} \subset \tau_{n_1,f_1} \cap \tau_{n_2,f_2}$$

since for $i = 1,2$ and $\mu \in \tau_{m,h}$

$$\hat{F}(f_i) - \mu(f_i) \leq \hat{F}(k_{n_i}) - \hat{F}(k_m) + \hat{F}(k_m) - \hat{F}(h) + \hat{F}(h) - \mu(h)$$
$$\leq \frac{1}{n_i} + \frac{1}{m} + \frac{3}{m} \leq \frac{3}{n_i} \quad .$$

Therefore,

$$\tau : = \{\tau_{n,f} : n \in \mathbb{N}, \quad f \in \mathcal{K}_+(E), \quad 1_K \leq f \leq k_n\}$$

is a filter base on M. Since M is compact, there is a filter Ψ on M finer than τ such that Ψ converges to an element $\mu_o \in M$. For $\varepsilon > 0$ there is $n \in \mathbb{N}$ such that $\frac{3}{n} \leq \varepsilon$ and a function $f \in \mathcal{K}_+(E)$ such that

$$1_K \leq f \leq k_n \quad \text{and}$$
$$\mu_o(f) < \mu_o(K) + \varepsilon \; .$$

Therefore, we have for some element $\Theta \in \Psi$

$$\mu(f) \leq \mu_o(K) + \varepsilon \quad \text{for all} \quad \mu \in \Theta \; .$$

Choose $\mu \in \Theta \cap \tau_{n,f} \in \Psi$. Then

$$|\mu_o(K) - F(K)| \leq |\mu_o(K) - \mu(f)| + |\mu(f) - \hat{F}(f)|$$
$$+ |\hat{F}(f) - F(K)| \leq \varepsilon + \frac{3}{n} + \frac{1}{n} \leq 3\varepsilon \; .$$

Since $\varepsilon > 0$ was arbitrary, we have $\mu_o(K) = F(K)$. $\mu_o \in M$ implies $\mu_o \leq F$ and

$$\|\mu_o\| \leq \|F\| = \sup_{n \in \mathbb{N}} \hat{F}(e_n) = \sup_{n \in \mathbb{N}} \mu_o(e_n) \leq \|\mu_o\|,$$

hence $\|\mu_o\| = \|F\|$.

<u>3.2 PROPOSITION</u>: Let F be a strongly subadditive capacity on E, let $U \subset E$ be open and relatively compact. Then $F_*(U) = \max\{\mu(U) : \mu \in \mathcal{M}_+(E), \mu \leq F, \|\mu\| = \|F\|\}$.

<u>PROOF</u>: By (2.1) for $K : = \emptyset$ and U there exist two sequences $(u_n)_{n \in \mathbb{N}}$ and $(e_n)_{n \in \mathbb{N}}$ in $\mathcal{K}_+(E)$ having the properties:

(1) $u_n \leq 1_U$, $e_n \leq 1$ for all $n \in \mathbb{N}$,

(2) $F_*(U) = \sup_{n \in \mathbb{N}} \hat{F}(u_n)$, $\|F\| = \sup_{n \in \mathbb{N}} \hat{F}(e_n)$,

(3) for $f : = 0$ there is a measure $\mu \in \mathcal{M}_+(E)$ such that $\mu \leq F$,

$$\mu(u_n) = \hat{F}(u_n), \quad \mu(e_n) = \hat{F}(e_n) \quad \text{for all} \quad n \in \mathbb{N}.$$

Therefore we have

$$\mu(U) \leq F_*(U) = \sup_{n \in \mathbb{N}} \hat{F}(u_n) = \sup_{n \in \mathbb{N}} \mu(u_n) \leq \mu(U) \qquad \text{and}$$

$$\|\mu\| \leq \|F\| = \sup_{n \in \mathbb{N}} \hat{F}(e_n) = \sup_{n \in \mathbb{N}} \mu(e_n) \leq \|\mu\| \quad , \qquad \text{hence}$$

$$\mu(U) = F_*(U) \quad \text{and} \quad \|\mu\| = \|F\|.$$

As a corollary of (3.1) we get

3.3 THEOREM: Let F be a strongly subadditive capacity on E and let $X \subset E$. Then

$$F_*(X) = \sup\,\{\mu_*(X) : \mu \in \mathcal{M}_+(E), \ \mu \leq F, \ \|\mu\| = \|F\|\}.$$

PROOF: Defining $\mathfrak{K}(X) := \{K \subset X : K \text{ compact}\}$ for $X \subset E$ and

$$M := \{\mu \in \mathcal{M}_+(E) : \mu \leq F, \ \|\mu\| = \|F\|\},$$

we get

$$F_*(X) = \sup_{K \in \mathfrak{K}(X)} F(K) = \sup_{K \in \mathfrak{K}(X)} \ \sup_{\mu \in M} \mu(K) = \sup_{\mu \in M} \ \sup_{K \in \mathfrak{K}(X)} \mu(K)$$

$$= \sup_{\mu \in M} \mu_*(X).$$

3.4 PROPOSITION: Let F be a strongly superadditive capacity on E. For any compact set $K \subset E$ we have

$$F(K) = \min\,\{\mu(K) : \mu \in \mathcal{M}_+(E), \ \mu \geq F, \ \|\mu\| = \|F\|\}.$$

PROOF: By (2.3), there exists a sequence $(k_n)_{n \in \mathbb{N}}$ in $\mathcal{K}_+(E)$ having the properties:

(1) $1_K \leq k_n$ for all $n \in \mathbb{N}$ and $F(K) = \inf_{n \in \mathbb{N}} \hat{F}(k_n)$,

(2) for $f := 0$ there is a measure $\mu \in \mathcal{M}_+(E)$ such that $\mu \geq F$, $\|\mu\| = \|F\|$ and $\mu(k_n) = \hat{F}(k_n)$ for all $n \in \mathbb{N}$.

Therefore,

$$\mu(K) \leq \inf_{n \in \mathbb{N}} \mu(k_n) = \inf_{n \in \mathbb{N}} \hat{F}(k_n) = F(K) \leq \mu(K), \quad \text{hence}$$

$$\mu(K) = F(K).$$

3.5 REMARK: If F is strongly superadditive and if

$\|F\|$ is infinite, the generalization of (3.4) is no longer true (even for open sets), as the following example shows:

Set $E: = \mathbb{R}$ and

$$F(K): = \begin{cases} \lambda(K), & 0 \in K \\ 0, & 0 \notin K \end{cases} \quad (K \in \mathcal{K}(\mathbb{R}), \quad \lambda \text{ denotes the Lebesgue measure}).$$

F is a strongly superadditive capacity on $\mathbb{R}$.

For $U: =]0, \infty[$ we have $F_*(U) = 0$.

$\mu \in \mathcal{M}_+(\mathbb{R})$ and $\mu \geq F$ imply

$\mu([0,n]) \geq F([0,n]) = \lambda([0,n]) = n$ and

$\mu(U) \geq \mu(]0,n]) = \mu([0,n]) - \mu(\{0\}) \geq n - \mu(\{0\})$

for all $n \in \mathbb{N}$, hence

$\mu(U) = \infty > 0 = F_*(U)$.

If $\|F\|$ is finite, we can prove the following result:

<u>3.6 PROPOSITION</u>: Let F be a strongly superadditive capacity on E such that $\|F\|$ is finite. If $U \subset E$ is open, then

$$F_*(U) = \min \{\mu(U) : \mu \in \mathcal{M}_+(E), \quad \mu \geq F, \quad \|\mu\| = \|F\|\}$$

<u>PROOF</u>: Let U be an open subset of E. By (1.4.5) there is an increasing sequence $(u_n)_{n\in\mathbb{N}}$ in $\mathcal{K}_+(E)$ such that

$$u_n \leq 1_U \quad \text{and} \quad F_*(U) - \hat{F}(u_n) \leq \frac{1}{n} \quad \text{for all} \quad n \in \mathbb{N}.$$

The set

$$\begin{aligned} M: &= \{\mu \in \mathcal{M}_+(E) : \mu \geq F, \|\mu\| = \|F\|\} \\ &= \{\mu \in \mathcal{M}_+(E) : \mu \geq F, \|\mu\| \leq \|F\|\} \end{aligned}$$

is a closed subset of

$$\prod_{f\in\mathcal{K}(E)} [-\|f\| \, \|F\|, \|f\| \, \|F\|]$$

and therefore compact in the vague topology.

For $n \in \mathbb{N}$ and $f \in \mathcal{K}_+(E)$ such that $u_n \leq f \leq 1_U$ set

$$\tau_{n,f} = \{\mu \in M : \mu(f) - \hat{F}(f) \leq \frac{3}{n}\}.$$

Then $\tau_{n,f} \neq \emptyset$ since by (2.3) applied to $K: = \operatorname{Supp} f$, there is a measure $\mu \in M$ such that $\mu(f) = \hat{F}(f)$.

For $i = 1,2$ let $n_i \in \mathbb{N}$ and $f_i \in \mathcal{K}_+(E)$ such that $u_{n_i} \leq f_i \leq 1_U$. Define $m: = 3 \sup (n_1,n_2)$ and $h: = \sup (u_m,f_1,f_2)$. Then $h \in \mathcal{K}_+(E)$ and $u_m \leq h \leq 1_U$. We have

$$\tau_{m,h} \subset \tau_{n_1,f_1} \cap \tau_{n_2,f_2}$$

since for $i = 1,2$ and $\mu \in \tau_{m,h}$ we obtain

$$\mu(f_i) - \hat{F}(f_i) \leq \mu(h) - \hat{F}(h) + \hat{F}(h) - \hat{F}(u_m) + \hat{F}(u_m) - \hat{F}(u_{n_i})$$
$$\leq \frac{3}{m} + \frac{1}{m} + \frac{1}{n_i} \leq \frac{3}{n_i} .$$

Therefore

$$\tau: = \{\tau_{n,f} : n \in \mathbb{N}, \quad f \in \mathcal{K}_+(E), \quad u_n \leq f \leq 1_U\}$$

is a filter base on M. Since M is compact, there is a filter Ψ on M finer than τ such that Ψ converges to an element $\mu_o \in M$. For $\varepsilon > 0$ there is $n \in \mathbb{N}$ such that $\frac{3}{n} \leq \varepsilon$ and a function $f \in \mathcal{K}_+(E)$ such that

$$u_n \leq f \leq 1_U \quad \text{and}$$
$$\mu_o(U) < \mu_o(f) + \varepsilon .$$

Therefore we have for some element $\Theta \in \Psi$

$$\mu_o(U) \leq \mu(f) + \varepsilon \quad \text{for all} \quad \mu \in \Theta .$$

Choose $\mu \in \Theta \cap \tau_{n,f} \in \Psi$. Then

$$|\mu_o(U) - F_*(U)| \leq |\mu_o(U) - \mu(f)| + |\mu(f) - \hat{F}(f)|$$
$$+ |\hat{F}(f) - F_*(U)| \leq \varepsilon + \frac{3}{n} + \frac{1}{n} \leq 3\varepsilon.$$

Since $\varepsilon > 0$ was arbitrary, we have $\mu_o(U) = F_*(U)$.

3.7 COROLLARY: Let F ba a strongly superadditive capacity on E such that $\|F\|$ is finite. Then for $X \subset E$

$$F^*(X) = \inf \{\mu^*(X) : \mu \in \mathcal{M}_+(E), \quad \mu \geq F, \quad \|\mu\| = \|F\|\}.$$

PROOF: Defining $\mathfrak{U}(X): = \{U \subset E \text{ open} : X \subset U\}$ and $M: = \{\mu \in \mathcal{M}_+(E) : \mu \geq F, \|\mu\| = \|F\|\}$,

we get by (3.6)

$$F^*(X) = \inf_{U \in \mathfrak{U}(X)} F_*(U) = \inf_{U \in \mathfrak{U}(X)} \inf_{\mu \in M} \mu(U) = \inf_{\mu \in M} \inf_{U \in \mathfrak{U}(X)} \mu(U)$$

$$= \inf_{\mu \in M} \mu^*(X).$$

3.8 REMARK: For an arbitrary capacity F, (3.3) or (3.6) are no longer true if F is not strongly subadditive or strongly superadditive, respectively, even if F^* is a capacity of Choquet in the sense of (1.2.3):

Consider the space $E := \{0,1\}$ with the discrete topology and define

$$F_1(\emptyset) = F_1(\{0\}) = F_1(\{1\}) = 0, \quad F_1(E) = 1,$$

$$F_2(\emptyset) = 0, \quad F_2(\{0\}) = F_2(\{1\}) = F_2(E) = 1.$$

Then F_1, F_2 are capacities of Choquet such that

$$F_i = (F_i)_* = F_i^* \text{ and } \|F_i\| = 1 \qquad (i = 1,2).$$

$\mu \in \mathcal{M}_+(E)$ and $\mu \leq F_1$ imply $\mu = 0$, hence

$$F_1(E) \neq \sup \{\mu(E) : \mu \in \mathcal{M}_+(E), \ \mu \leq F_1\}.$$

$\mu \in \mathcal{M}_+(E)$ and $\mu \geq F_2$ imply

$\mu(E) \geq 2$, hence

$$F_2(E) \neq \inf \{\mu(E) : \mu \in \mathcal{M}_+(E), \ \mu \geq F_2\}.$$

Bibliography

H.BAUER [1] : Über die Fortsetzung positiver Linearformen. Ber. Bayer. Akad. Wiss., math.-naturw. Kl. München 1957, 177 - 190.

G.CHOQUET [2] : Theory of Capacities. Ann. Inst. Fourier 5, 131 - 295 (1953/54).

[3] : Lectures on Analysis. New York - Amsterdam: Benjamin 1969.

C.DELLACHERIE [4] : Quelques commentaires sur les prolongements de capacités. Lect. Notes Math. 191 (Sém. de Prob. V), 77 - 81 (1971).

V.STRASSEN [5] : Meßfehler und Information. Z.Wahrscheinlichkeitstheorie 2, 273 - 305 (1964).

Contents

Lecture Notes in Mathematics

Comprehensive leaflet on request

Vol. 38: R. Berger, R. Kiehl, E. Kunz und H.-J. Nastold, Differentialrechnung in der analytischen Geometrie IV, 134 Seiten. 1967 DM 12,-

Vol. 39: Séminaire de Probabilités I. II, 189 pages. 1967. DM 14,-

Vol. 40: J. Tits, Tabellen zu den einfachen Lie Gruppen und ihren Darstellungen. VI, 53 Seiten. 1967. DM 6,80

Vol. 41: A. Grothendieck, Local Cohomology. VI, 106 pages. 1967. DM 10,-

Vol. 42: J. F. Berglund and K. H. Hofmann, Compact Semitopological Semigroups and Weakly Almost Periodic Functions. VI, 160 pages. 1967. DM 12,-

Vol. 43: D. G. Quillen, Homotopical Algebra. VI, 157 pages. 1967. DM 14,-

Vol. 44: K. Urbanik, Lectures on Prediction Theory. IV, 50 pages. 1967. DM 5,80

Vol. 45: A. Wilansky, Topics in Functional Analysis. VI, 102 pages. 1967. DM 9,60

Vol. 46: P. E. Conner, Seminar on Periodic Maps. IV, 116 pages. 1967. DM 10,60

Vol. 47: Reports of the Midwest Category Seminar I. IV, 181 pages. 1967. DM 14,80

Vol. 48: G. de Rham, S. Maumary et M. A. Kervaire, Torsion et Type Simple d'Homotopie. IV, 101 pages. 1967. DM 9,60

Vol. 49: C. Faith, Lectures on Injective Modules and Quotient Rings. XVI, 140 pages. 1967. DM 12,80

Vol. 50: L. Zalcman, Analytic Capacity and Rational Approximation. VI, 155 pages. 1968. DM 13,20

Vol. 51: Séminaire de Probabilités II. IV, 199 pages. 1968. DM 14,-

Vol. 52: D. J. Simms, Lie Groups and Quantum Mechanics. IV, 90 pages. 1968. DM 8,-

Vol. 53: J. Cerf, Sur les difféomorphismes de la sphère de dimension trois ($\Gamma_4 = O$). XII, 133 pages. 1968. DM 12,-

Vol. 54: G. Shimura, Automorphic Functions and Number Theory. VI, 69 pages. 1968. DM 8,-

Vol. 55: D. Gromoll, W. Klingenberg und W. Meyer, Riemannsche Geometrie im Großen. VI, 287 Seiten. 1968. DM 20,-

Vol. 56: K. Floret und J. Wloka, Einführung in die Theorie der lokalkonvexen Räume. VIII, 194 Seiten. 1968. DM 16,-

Vol. 57: F. Hirzebruch und K. H. Mayer, O (n)-Mannigfaltigkeiten, exotische Sphären und Singularitäten. IV, 132 Seiten. 1968. DM 10,80

Vol. 58: Kuramochi Boundaries of Riemann Surfaces. IV, 102 pages. 1968. DM 9,60

Vol. 59: K. Jänich, Differenzierbare G-Mannigfaltigkeiten. VI, 89 Seiten. 1968. DM 8,-

Vol. 60: Seminar on Differential Equations and Dynamical Systems. Edited by G. S. Jones. VI, 106 pages. 1968. DM 9,60

Vol. 61: Reports of the Midwest Category Seminar II. IV, 91 pages. 1968. DM 9,60

Vol. 62: Harish-Chandra, Automorphic Forms on Semisimple Lie Groups X, 138 pages. 1968. DM 14,-

Vol. 63: F. Albrecht, Topics in Control Theory. IV, 65 pages. 1968. DM 6,80

Vol. 64: H. Berens, Interpolationsmethoden zur Behandlung von Approximationsprozessen auf Banachräumen. VI, 90 Seiten. 1968. DM 8,-

Vol. 65: D. Kölzow, Differentiation von Maßen. XII, 102 Seiten. 1968. DM 8,-

Vol. 66: D. Ferus, Totale Absolutkrümmung in Differentialgeometrie und -topologie. VI, 85 Seiten. 1968. DM 8,-

Vol. 67: F. Kamber and P. Tondeur, Flat Manifolds. IV, 53 pages. 1968. DM 5,80

Vol. 68: N. Boboc et P. Mustată, Espaces harmoniques associés aux opérateurs différentiels linéaires du second ordre de type elliptique. VI, 95 pages. 1968. DM 8,60

Vol. 69: Seminar über Potentialtheorie. Herausgegeben von H. Bauer. VI, 180 Seiten. 1968. DM 14,80

Vol. 70: Proceedings of the Summer School in Logic. Edited by M. H. Löb. IV, 331 pages. 1968. DM 20,-

Vol. 71: Séminaire Pierre Lelong (Analyse), Année 1967 - 1968. VI, 190 pages. 1968. DM 14,

Vol. 72: The Syntax and Semantics of Infinitary Languages. Edited by J. Barwise. IV, 268 pages. 1968. DM 18,-

Vol. 73: P. E. Conner, Lectures on the Action of a Finite Group. IV, 123 pages. 1968. DM 10,-

Vol. 74: A. Fröhlich, Formal Groups. IV, 140 pages. 1968. DM 12,-

Vol. 75: G. Lumer, Algèbres de fonctions et espaces de Hardy. VI, 80 pages. 1968. DM 8,

Vol. 76: R. G. Swan, Algebraic K-Theory. IV, 262 pages. 1968. DM 18,-

Vol. 77: P.-A. Meyer, Processus de Markov: la frontière de Martin. IV, 123 pages. 1968. DM 10,-

Vol. 78: H. Herrlich, Topologische Reflexionen und Coreflexionen. XVI, 166 Seiten. 1968. DM 12,-

Vol. 79: A. Grothendieck, Catégories Cofibrées Additives et Complexe Cotangent Relatif. IV, 167 pages. 1968. DM 12,-

Vol. 80: Seminar on Triples and Categorical Homology Theory. Edited by B. Eckmann. IV, 398 pages. 1969. DM 20,-

Vol. 81: J.-P. Eckmann et M. Guenin, Méthodes Algébriques en Mécanique Statistique. VI, 131 pages. 1969. DM 12,-

Vol. 82: J. Wloka, Grundräume und verallgemeinerte Funktionen. VIII, 131 Seiten. 1969. DM 12,-

Vol. 83: O. Zariski, An Introduction to the Theory of Algebraic Surfaces. IV, 100 pages. 1969. DM 8,-

Vol. 84: H. Lüneburg, Transitive Erweiterungen endlicher Permutationsgruppen. IV, 119 Seiten. 1969. DM 10,-

Vol. 85: P. Cartier et D. Foata, Problèmes combinatoires de commutation et réarrangements. IV, 88 pages. 1969. DM 8,-

Vol. 86: Category Theory, Homology Theory and their Applications I. Edited by P. Hilton. VI, 216 pages. 1969. DM 16,-

Vol. 87: M. Tierney, Categorical Constructions in Stable Homotopy Theory. IV, 65 pages. 1969. DM 6,-

Vol. 88: Séminaire de Probabilités III. IV, 229 pages. 1969. DM 18,-

Vol. 89: Probability and Information Theory. Edited by M. Behara, K. Krickeberg and J. Wolfowitz. IV, 256 pages. 1969. DM 18,-

Vol. 90: N. P. Bhatia and O. Hajek, Local Semi-Dynamical Systems. II, 157 pages. 1969. DM 14,-

Vol. 91: N. N. Janenko, Die Zwischenschrittmethode zur Lösung mehrdimensionaler Probleme der mathematischen Physik. VIII, 194 Seiten. 1969. DM 16,80

Vol. 92: Category Theory, Homology Theory and their Applications II. Edited by P. Hilton. V, 308 pages. 1969. DM 20,-

Vol. 93: K. R. Parthasarathy, Multipliers on Locally Compact Groups. III, 54 pages. 1969. DM 5,60

Vol. 94: M. Machover and J. Hirschfeld, Lectures on Non-Standard Analysis. VI, 79 pages. 1969. DM 6,-

Vol. 95: A. S. Troelstra, Principles of Intuitionism. II, 111 pages. 1969. DM 10,-

Vol. 96: H.-B. Brinkmann und D. Puppe, Abelsche und exakte Kategorien, Korrespondenzen. V, 141 Seiten. 1969. DM 10,-

Vol. 97: S. O. Chase and M. E. Sweedler, Hopf Algebras and Galois theory. II, 133 pages. 1969. DM 10,-

Vol. 98: M. Heins, Hardy Classes on Riemann Surfaces. III, 106 pages. 1969. DM 10,-

Vol. 99: Category Theory, Homology Theory and their Applications III. Edited by P. Hilton. IV, 489 pages. 1969. DM 24,-

Vol. 100: M. Artin and B. Mazur, Etale Homotopy. II, 196 Seiten. 1969. DM 12,-

Vol. 101: G. P. Szegö et G. Treccani, Semigruppi di Trasformazioni Multivoche. VI, 177 pages. 1969. DM 14,-

Vol. 102: F. Stummel, Rand- und Eigenwertaufgaben in Sobolewschen Räumen. VIII, 386 Seiten. 1969. DM 20,-

Vol. 103: Lectures in Modern Analysis and Applications I. Edited by C. T. Taam. VII, 162 pages. 1969. DM 12,-

Vol. 104: G. H. Pimbley, Jr., Eigenfunction Branches of Nonlinear Operators and their Bifurcations. II, 128 pages. 1969. DM 10,-

Vol. 105: R. Larsen, The Multiplier Problem. VII, 284 pages. 1969. DM 18,-

Vol. 106: Reports of the Midwest Category Seminar III. Edited by S. Mac Lane. III, 247 pages. 1969. DM 16,-

Vol. 107: A. Peyerimhoff, Lectures on Summability. III, 111 pages. 1969. DM 8,-

Vol. 108: Algebraic K-Theory and its Geometric Applications. Edited by R. M. F. Moss and C. B. Thomas. IV, 86 pages. 1969. DM 6,-

Vol. 109: Conference on the Numerical Solution of Differential Equations. Edited by J. Ll. Morris. VI, 275 pages. 1969. DM 18,-

Vol. 110: The Many Facets of Graph Theory. Edited by G. Chartrand and S. F. Kapoor. VIII, 290 pages. 1969. DM 18,-

Vol. 111: K. H. Mayer, Relationen zwischen charakteristischen Zahlen. III, 99 Seiten. 1969. DM 8,-

Vol. 112: Colloquium on Methods of Optimization. Edited by N. N. Moiseev. IV, 293 pages. 1970. DM 18,-

Vol. 113: R. Wille, Kongruenzklassengeometrien. III, 99 Seiten. 1970. DM 8,-

Vol. 114: H. Jacquet and R. P. Langlands, Automorphic Forms on GL (2). VII, 548 pages. 1970. DM 24,-

Vol. 115: K. H. Roggenkamp and V. Huber-Dyson, Lattices over Orders I. XIX, 290 pages. 1970. DM 18,-

Vol. 116: Séminaire Pierre Lelong (Analyse) Année 1969. IV, 195 pages. 1970. DM 14,-

Vol. 117: Y. Meyer, Nombres de Pisot, Nombres de Salem et Analyse Harmonique. 63 pages. 1970. DM 6,-

Vol. 118: Proceedings of the 15th Scandinavian Congress, Oslo 1968. Edited by K. E. Aubert and W. Ljunggren. IV, 162 pages. 1970. DM 12,-

Vol. 119: M. Raynaud, Faisceaux amples sur les schémas en groupes et les espaces homogènes. III, 219 pages. 1970. DM 14,-

Vol. 120: D. Siefkes, Büchi's Monadic Second Order Successor Arithmetic. XII, 130 Seiten. 1970. DM 12,-

Vol. 121: H. S. Bear, Lectures on Gleason Parts. III, 47 pages. 1970. DM 6,-

Vol. 122: H. Zieschang, E. Vogt und H.-D. Coldewey, Flächen und ebene diskontinuierliche Gruppen. VIII, 203 Seiten. 1970. DM 16,-

Vol. 123: A. V. Jategaonkar, Left Principal Ideal Rings. VI, 145 pages. 1970. DM 12,-

Vol. 124: Séminare de Probabilités IV. Edited by P. A. Meyer. IV, 282 pages. 1970. DM 20,-

Vol. 125: Symposium on Automatic Demonstration. V, 310 pages. 1970. DM 20,-

Vol. 126: P. Schapira, Théorie des Hyperfonctions. XI, 157 pages. 1970. DM 14,-

Vol. 127: I. Stewart, Lie Algebras. IV, 97 pages. 1970. DM 10,-

Vol. 128: M. Takesaki, Tomita's Theory of Modular Hilbert Algebras and its Applications. II, 123 pages. 1970. DM 10,-

Vol. 129: K. H. Hofmann, The Duality of Compact Semigroups and C*- Bigebras. XII, 142 pages. 1970. DM 14,-

Vol. 130: F. Lorenz, Quadratische Formen über Körpern. II, 77 Seiten. 1970. DM 8,-

Vol. 131: A Borel et al., Seminar on Algebraic Groups and Related Finite Groups. VII, 321 pages. 1970. DM 22,

Vol. 132: Symposium on Optimization. III, 348 pages. 1970. DM 22,-

Vol. 133: F. Topsøe, Topology and Measure. XIV, 79 pages. 1970. DM 8,-

Vol. 134: L. Smith, Lectures on the Eilenberg-Moore Spectral Sequence. VII, 142 pages. 1970. DM 14,-

Vol. 135: W. Stoll, Value Distribution of Holomorphic Maps into Compact Complex Manifolds. II, 267 pages. 1970. DM 18,-

Vol. 136: M. Karoubi et al., Séminaire Heidelberg-Saarbrücken-Strasbourg sur la K-Théorie. IV, 264 pages. 1970. DM 18,-

Vol. 137: Reports of the Midwest Category Seminar IV. Edited by S. MacLane. III, 139 pages. 1970. DM 12,-

Vol. 138: D. Foata et M. Schützenberger, Théorie Géométrique des Polynômes Eulériens. V, 94 pages. 1970. DM 10,-

Vol. 139: A. Badrikian, Séminaire sur les Fonctions Aléatoires Linéaires et les Mesures Cylindriques. VII, 221 pages. 1970. DM 18,-

Vol. 140: Lectures in Modern Analysis and Applications II. Edited by C. T. Taam. VI, 119 pages. 1970. DM 10,-

Vol. 141: G. Jameson, Ordered Linear Spaces. XV, 194 pages. 1970. DM 16,-

Vol. 142: K. W. Roggenkamp, Lattices over Orders II. V, 388 pages. 1970. DM 22,-

Vol. 143: K. W. Gruenberg, Cohomological Topics in Group Theory. XIV, 275 pages. 1970. DM 20,-

Vol. 144: Seminar on Differential Equations and Dynamical Systems, II. Edited by J. A. Yorke. VIII, 268 pages. 1970. DM 20,-

Vol. 145: E. J. Dubuc, Kan Extensions in Enriched Category Theory. XVI, 173 pages. 1970. DM 16,-

Vol. 146: A. B. Altman and S. Kleiman, Introduction to Grothendieck Duality Theory. II, 192 pages. 1970. DM 18,-

Vol. 147: D. E. Dobbs, Cech Cohomological Dimensions for Commutative Rings. VI, 176 pages. 1970. DM 16,-

Vol. 148: R. Azencott, Espaces de Poisson des Groupes Localement Compacts. IX, 141 pages. 1970. DM 14,-

Vol. 149: R. G. Swan and E. G. Evans, K-Theory of Finite Groups and Orders. IV, 237 pages. 1970. DM 20,-

Vol. 150: Heyer, Dualität lokalkompakter Gruppen. XIII, 372 Seiten. 1970. DM 20,-

Vol. 151: M. Demazure et A. Grothendieck, Schémas en Groupes I. (SGA 3). XV, 562 pages. 1970. DM 24,-

Vol. 152: M. Demazure et A. Grothendieck, Schémas en Groupes II. (SGA 3). IX, 654 pages. 1970. DM 24,-

Vol. 153: M. Demazure et A. Grothendieck, Schémas en Groupes III. (SGA 3). VIII, 529 pages. 1970. DM 24,-

Vol. 154: A. Lascoux et M. Berger, Variétés Kähleriennes Compactes. VII, 83 pages. 1970. DM 8,-

Vol. 155: Several Complex Variables I, Maryland 1970. Edited by J. Horváth. IV, 214 pages. 1970. DM 18,-

Vol. 156: R. Hartshorne, Ample Subvarieties of Algebraic Varieties. XIV, 256 pages. 1970. DM 20,-

Vol. 157: T. tom Dieck, K. H. Kamps und D. Puppe, Homotopietheorie. VI, 265 Seiten. 1970. DM 20,-

Vol. 158: T. G. Ostrom, Finite Translation Planes. IV. 112 pages. 1970. DM 10,-

Vol. 159: R. Ansorge und R. Hass. Konvergenz von Differenzenverfahren für lineare und nichtlineare Anfangswertaufgaben. VIII, 145 Seiten. 1970. DM 14,-

Vol. 160: L. Sucheston, Constributions to Ergodic Theory and Probability. VII, 277 pages. 1970. DM 20,-

Vol. 161: J. Stasheff, H-Spaces from a Homotopy Point of View. VI, 95 pages. 1970. DM 10,-

Vol. 162: Harish-Chandra and van Dijk, Harmonic Analysis on Reductive p-adic Groups. IV, 125 pages. 1970. DM 12,-

Vol. 163: P. Deligne, Equations Différentielles à Points Singuliers Reguliers. III, 133 pages. 1970. DM 12,-

Vol. 164: J. P. Ferrier, Seminaire sur les Algebres Complétes. II, 69 pages. 1970. DM 8,-

Vol. 165: J. M. Cohen, Stable Homotopy. V, 194 pages. 1970. DM 16,-

Vol. 166: A. J. Silberger, PGL_2 over the p-adics: its Representations, Spherical Functions, and Fourier Analysis. VII, 202 pages. 1970. DM 18,-

Vol. 167: Lavrentiev, Romanov and Vasiliev, Multidimensional Inverse Problems for Differential Equations. V, 59 pages. 1970. DM 10,-

Vol. 168: F. P. Peterson, The Steenrod Algebra and its Applications: A conference to Celebrate N. E. Steenrod's Sixtieth Birthday. VII, 317 pages. 1970. DM 22,-

Vol. 169: M. Raynaud, Anneaux Locaux Henséliens. V, 129 pages. 1970. DM 12,-

Vol. 170: Lectures in Modern Analysis and Applications III. Edited by C. T. Taam. VI, 213 pages. 1970. DM 18,-

Vol. 171: Set-Valued Mappings, Selections and Topological Properties of 2^X. Edited by W. M. Fleischman. X, 110 pages. 1970. DM 12,-

Vol. 172: Y.-T. Siu and G. Trautmann, Gap-Sheaves and Extension of Coherent Analytic Subsheaves. V, 172 pages. 1971. DM 16,-

Vol. 173: J. N. Mordeson and B. Vinograde, Structure of Arbitrary Purely Inseparable Extension Fields. IV, 138 pages. 1970. DM 14,-

Vol. 174: B. Iversen, Linear Determinants with Applications to the Picard Scheme of a Family of Algebraic Curves. VI, 69 pages. 1970. DM 8,-

Vol. 175: M. Brelot, On Topologies and Boundaries in Potential Theory. VI, 176 pages. 1971. DM 18,-

Vol. 176: H. Popp, Fundamentalgruppen algebraischer Mannigfaltigkeiten. IV, 154 Seiten. 1970. DM 16,-

Vol. 177: J. Lambek, Torsion Theories, Additive Semantics and Rings of Quotients. VI, 94 pages. 1971. DM 12,-

Vol. 178: Th. Bröcker und T. tom Dieck, Kobordismentheorie. XVI, 191 Seiten. 1970. DM 18,-

Vol. 179: Seminaire Bourbaki – vol. 1968/69. Exposés 347-363. IV, 295 pages. 1971. DM 22,-

Vol. 180: Séminaire Bourbaki – vol. 1969/70. Exposés 364-381. IV, 310 pages. 1971. DM 22,-

Vol. 181: F. DeMeyer and E. Ingraham, Separable Algebras over Commutative Rings. V, 157 pages. 1971. DM 16,-